❖ ❖ ❖ ❖ ❖ ❖ ❖ ❖ ❖ ❖ ❖

HUMAN DIVERSITY
A Guide for Understanding
Third Edition

Stuart E. Schwartz
University of Florida

Craig A. Conley

The McGraw-Hill Companies, Inc.
Primis Custom Publishing

New York St. Louis San Francisco Auckland Bogotá
Caracas Lisbon London Madrid Mexico Milan Montreal
New Delhi Paris San Juan Singapore Sydney Tokyo Toronto

McGraw-Hill

A Division of The ***McGraw-Hill*** *Companies*

HUMAN DIVERSITY

A Guide for Understanding

Models: John Albritten, Justin Best, Jacqueline Bush, Brianna Corrigan, Chris George, Natasha Herard, Jeffrey S. Kissinger, Elissa Kleinman, Frances C. Langford, Alex Logiou, Robert Malik Lawrence, Jennifer C. Meyers, Lakisha Mobley, Jose A. Perez, Ann Elizabeth Perry, Kelly-Lynne Scalice, Stacy Schoenherr, Paul Steele.

McGraw-Hill's Primis Custom Publishing consists of products that are produced from camera-ready copy. Peer review, class testing, and accuracy are primarily the responsibility of the author(s).

2 3 4 5 6 7 8 9 0 DOC DOC 0 9

ISBN 0-07-058000-6

Editor: Judy T. Ice

Interior Design and Layout: Craig A. Conley

Photography: Frank Conley, Kurt Lischka, Louis Mallory, Michael Weimar

Sign Language Drawings: Robert Stiff

Cover Designer: Maggie Lytle

Printer/Binder: R.R. Donnelley & Sons Company

For Rita, Charles, and George Schaefer

—SES

For Cassie Farley

—CAC

Acknowledgements

Special thanks from SES to:

The many dedicated persons who have served as graduate and undergraduate teaching assistants in Exceptional People. You have been a constant inspiration and source of knowledge.

George Schaefer who has provided motivation and reinforcement for this project.

Special thanks from CAC to:

June Conley for her invaluable insights, suggestions, contributions, and improvements. Your vast knowledge and wisdom are unparalleled.

Michael Warwick for his multifaceted assistance and support with this project.

Special thanks from both:

To Belinda Karge and Ravic Ringlaben for their work on the instructor's activity guide which accompanies this text.

To Lisa Eaton for her work on the Exceptional People correspondence course.

Our sincerest thanks go to Judy Ice, our McGraw Hill Editor. She has served as a motivator, teacher, listener, manager, and friend. Judy is a true professional who knows how to guide the content, people, and business aspects of a book development project.

SES & CAC

Preface

The writing of a textbook is not an easy task. However the recognition that this book may assist individuals in their understanding of human diversity and help improve their ability to interact with and respect others who are different from themselves is outstanding motivation. I teach courses about human diversity at the University of Florida and have frequent opportunities to hear from my graduates. Letters from former students which inform me that they are now better able to cope with a child who has a disability, that they are comfortable with new friends who have a different sexual orientation, that they have been influential in the hiring of a diverse individual, or that they have chosen a career related to working with persons who are diverse are the true rewards of teaching our courses. This book extends my opportunity to be helpful to other students who, like my own students, have enrolled in a course in order to learn about diversity in society.

My co-author and I suggest that you begin your study by becoming familiar with the components of the book. You'll find that the book is organized into three sections. In the first section you will be introduced to general issues regarding persons who are diverse and provide you with an overview of the culture of diversity. In Section One you will also find a review of significant legislation which affects individuals who are diverse.

Section Two discusses specific categories of human diversity. Within each of the topics, which we recognize could easily by themselves be developed to such a level of depth that a book would be needed, we have given definitions, causes, terminology, and have suggested appropriate interactions. It is our hope that this information will assist you in reducing myths and misconceptions and give you suggestions for comfortable interactions with persons who are different from you.

Section Three starts with a discussion of issues relating to parents, siblings, and other family members and their interactions with those in the family who are different. Next we provide an overview of educational perspectives regarding diverse individuals. This discussion should assist you if you are a parent of a child who needs special education services or who may experience educational difficulties due to being different from typical classmates. It may also provide you with an introductory knowledge which is crucial if you are going into the field of special education or a related service. The final chapter is intended to challenge you to review your attitude, your level of respect for others, and your tolerance for disrespect toward diverse individuals by friends and colleagues.

Be sure to note that there are many components of the book which are designed to assist you. At the end of each chapter there are activities which we strongly encourage you to complete. A study guide with sample test questions is also provided at the end of the book. It is clear, from data collection and analysis, that students who have completed these activities have scored higher on course tests; we hope you will use these activities to your advantage.

We want each of you to profit from your study of human diversity. If you started with some level of discomfort or misunderstanding about those who are different from you, that's perfectly understandable. All we want is for you to use this book and the course you are taking as an opportunity to assess your individual comfort level with those who are different. Recognize that each of us has some type of difference and we usually accept and respect those differences we are used to. Perhaps your study of human diversity will enable you to become used to other types of differences and therefore improve your comfort level in respect to others who are different.

I would love to hear from you. Your reactions and thoughts to your study of human diversity are very helpful and important to me. Please feel free to e-mail or write to me. Enjoy your study of human diversity!

Stuart E. Schwartz, Ed.D.
Box 117050
University of Florida
Gainesville FL 32611
voice: (352) 392-0701 ext. 258
e-mail: ses@coe.ufl.edu
homepage: www.coe.ufl.edu/DiverseWorld/

About the Authors

Stuart E. Schwartz

Stuart E. Schwartz joined the University of Florida special education faculty in 1974 after obtaining his doctoral degree from the University of Kansas in the area of transition for students with disabilities. He taught in the Philadelphia area, prior to his graduate studies, in middle school and high school special education, and he coordinated a work study program for high school students with disabilities.

Dr. Schwartz has published numerous articles regarding research in transition and has developed five curriculum materials which are currently being used in high school special education programs. He is the author of *Coping with Crisis Situations in the Classroom* and has presented at many national conferences on the topics of human diversity, persons with disabilities, transition services, and sexual orientation.

Dr. Schwartz regularly teaches human diversity courses for more than 2100 non education majors each year. He has been selected as "Teacher of the Year" twice since 1985.

Craig A. Conley

Craig A. Conley holds a B.S. in Mass Communications and an M.A. in English from Middle Tennessee State University. For nine years he was an instructor of composition, literature, and study skills at universities and community colleges in Tennessee, Virginia, and Florida. He is currently a consulting editor for McGraw-Hill and Globe-Fearon publishers.

Conley is co-author of *Diversity in the Classroom, Diversity on Campus: An Opportunity for Reflection,* and *Human Diversity: A Visual Resource,* published by McGraw-Hill. Four of his essays have been anthologized in the books *Alternatives in Education* and *The Homeschool Reader,* published by Home Education Press. His first novel, *Suspended Disbelief,* is published by Commonwealth.

Contents

Section I:

An Overview of Diversity

Chapter 1 Facets of Human Diversity

Chapter Sections

- Introduction
- What Constitutes Exceptionality?
- Historical Perspectives
- Terms and Labels
- Attitudes
- Conclusion
- Individual and Group Activities

"No loss by flood and lightning, no destruction of cities and temples by hostile forces of nature, has deprived man of so many noble lives and impulses as those which his intolerance has destroyed."
—Helen Keller, educator

Objectives

- Explain why people tend to subdivide.
- Describe what constitutes exceptionality.
- Discuss how attitudes about human diversity have changed over time.
- Identify the three crucial criteria of labels we use.

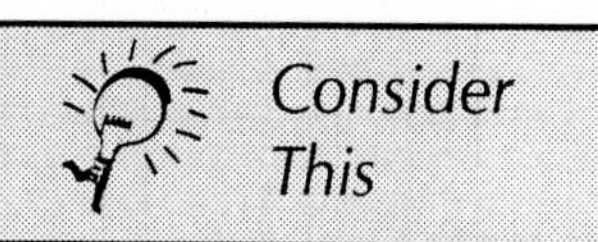

Imagine yourself as a person living in a world where everything is suited for people who are at least eight feet tall. How are you going to climb a flight of stairs, when each stair comes up to your waist? What would be easier to deal with—the stairs or the stares? Why?

Introduction

Life in the world is fast and busy, and too often people act on their habit of hate, fear, and ignorant reaction to anything perceived as different. Such reactions put up barriers to success, cutting people off from the richness of human diversity all around them. Look around you and observe ten people. Using statistics from the last Department of Education census report, two of those people will be foreign born, two will have a speech problem, and one will be a member of a minority religious group. If your random sampling includes persons with children, two children will have learning disabilities, two will face future drug or alcohol abuse, one will be developmentally delayed, and one will be homosexual.

Just as no two diamonds are exactly alike, every individual is unique (see Figure 1.1). But in our daily interactions with others we like to sub-divide sometimes, feeling that we belong in the same category as some, and not as others. We may tend to focus on the ways in which we are different because it makes us feel secure as part of a small group or societal designation, or because we feel uncomfortable or awkward with people who are different. Whenever our reactions to another are based on gender, race, physical or mental disability, sexual orientation, religion, economic class, or any other category, we are cheating ourselves of opportunities to

Figure 1.1
A diamond has many facets, each sparkling in its own way. In order to see the whole diamond, you must look at all the facets. Every diamond and every person is unique. To appreciate other members of the community, look at the different things that make them shine.

Personal Checklist

- ☑ I have hopes.
- ❑ I have dreams.
- ❑ I have experiences.
- ❑ I have memories.
- ❑ Sometimes I worry.
- ❑ Sometimes I feel afraid.
- ❑ I do some things well.
- ❑ I have an age.
- ❑ I have a size.
- ❑ I have a gender.
- ❑ I have a race.
- ❑ I have moods.
- ❑ I have a skin color.
- ❑ I have a nationality.
- ❑ I have beliefs.
- ❑ I have a culture.
- ❑ I have a heritage.
- ❑ I learn in my own way.
- ❑ I have a language.
- ❑ I have a personality.
- ❑ I act in my own way.
- ❑ I have senses.

Figure 1.2
Every person has special talents and skills, different experiences, original ideas, personal views and beliefs, and different feelings and emotions.

benefit from that individual and we are cheating that person of his or her right to participate fully in the human community.

Most of us have found ourselves uncomfortable at one time or another in dealing with someone we perceive to be different. For example, we may feel unsure of how to talk to someone from another culture, we may have incorrect expectations regarding an individual with a different skin color, we may have difficulty having a relaxed conversation with a person of a different sexual orientation, or we may find ourselves staring at someone who looks different due to an injury or a disability. In order to enhance our interactions with others and to increase our self-understanding, we need to know what human differences are, examine some of their causes, and reduce our misunderstandings and myths about people who are diverse. Since every one of us fits into some category of diversity (see Figure 1.2), we severely impoverish our lives if we limit our interaction through ignorance.

This book should help you to understand exactly how to define human differences and give you some guidance in regard to social interaction with various people. It includes helpful exercises and activities which will give you insights into other perspectives and will help you to develop sensitivity and self-awareness so that you can treat all people respectfully and fairly. Knowledge allows us to live and respond based on facts rather than myths. Information is the key to interacting with others with tolerance and respect, which is a challenge most people want to meet.

In order to profit best from this course it is advisable to first acquaint yourself with this text. Each chapter provides insightful discussions for your consideration. The book explores the effects of being different both upon the person and upon his or her family. You will be introduced to the concept of diversity, learn a variety of terms, and review some myths and misconceptions about diverse people.

What Constitutes Exceptionality?

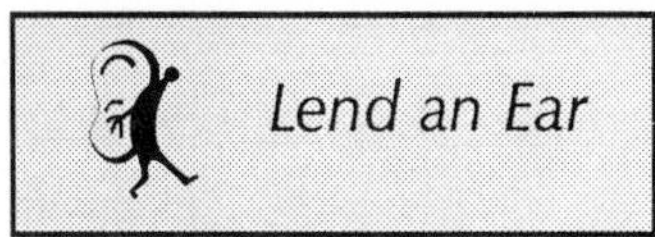

"The art of being wise is the art of knowing what to overlook." —William James, philosopher

Exceptional individuals—often called "persons who are diverse" or "persons who are different" throughout this book—significantly differ, in one or more ways, from those who are considered to be average. This definition could, of course, apply to anyone, depending upon the group to which he or she is being compared. Historically, Caucasian, heterosexual, English speakers have not been considered exceptional in the United States. However, in some parts of our country now, Spanish is the predominant language, so English speakers are the ones who are different. In many communities Hispanic or African American families outnumber Caucasian families, so it is exceptional to be white there. In a group of mostly females, it is exceptional to be male.

The expectations that a certain population has upon members of its group for specific behaviors or group standards also affect whether the individual is considered different. A group of teens which expects its friends to dress in a specific style might consider anyone who dresses differently to be exceptional. Parents who have high academic expectations for their child might consider C's to be exceptional grades, while other parents might find those grades "average" and acceptable. How we are viewed and accepted depends largely upon the situation in which we find ourselves (see Figure 1.3).

Some diverse people have disabilities which inhibit or prevent their participation in some activities or interfere with their learning. Some have special gifts or talents which make them different. And some have both talents and disabilities. Although diverse people may differ from others in some major or minor way, they are individuals first. Their exceptionality is only one of their characteristics. They each have their own unique goals, dreams, hopes, and needs.

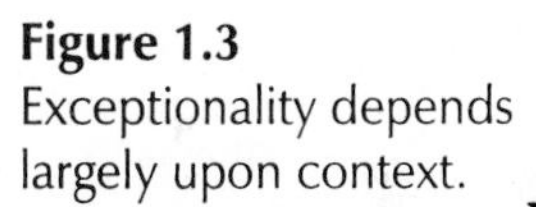

Figure 1.3
Exceptionality depends largely upon context.

Many of us may think of disabilities as physical limitations. Indeed, a disability may be physical, but it may also be emotional, cognitive, or sensory. A disability may cause one to use a wheelchair, or it may be only a minor handicap or no handicap at all. For instance, most people with vision or hearing impairments can easily correct their conditions with eyeglasses or hearing aids. You can see that diverse individuals function in every part of society every day.

In order to speak about human diversity with any real meaning, then, we would need to consider every possible difference from the norm, every possible norm, and every possible level of expectation as we examine the concept of exceptionality. That is clearly impractical if not impossible, but we can look at the most common types of diversity which you are likely to encounter in your school, your family, and your place of employment. These selected categories should enable you to have a thorough understanding of these areas of diversity and should help you to be a better friend, neighbor, family member, or employer of people who may, from time to time or from group to group, be considered different.

You yourself may now or in the future belong to a group which is characterized as different because of your race, gender, sexual orientation, behavioral style, functional limitation, size, age, religion, language, health, or socio-economic class. An understanding of the fact that the perception of diversity shifts constantly depending upon time and place will help you to avoid the insensitivity that comes from ignorance (see Figure 1.4). This book should help you, as well, to have a better understanding of yourself as an diverse person in our complex society.

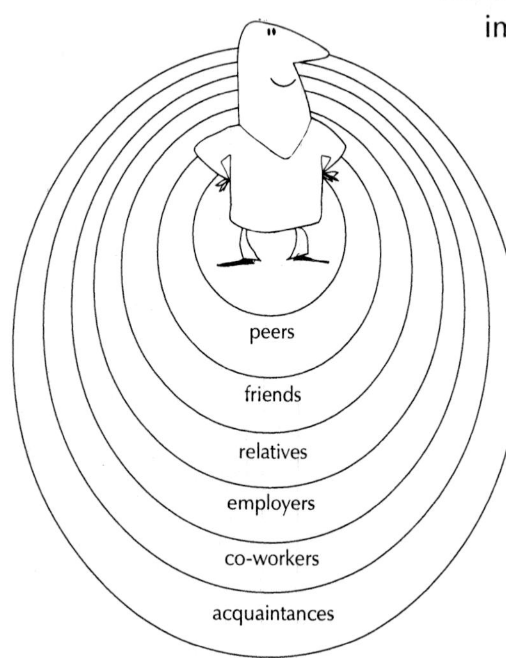

Figure 1.4
We encounter diversity in various social circles, finding ourselves exceptional in different ways in different contexts.

Historical Perspectives

Each category of diversity has, of course, had its own unique history. We can, however, speak in broad terms about some aspects of the acceptance and treatment of exceptional people in the past. For example, we know that the teachings of all the world's major religions include references to those who are sick, weak, unfortunate, or disadvantaged. Confucius, Buddha, Mohammed and Jesus all wrote or spoke about the need to show compassion. The sacred texts of the great religions all advocate humane and sympathetic treatment for every individual (see Figure 1.5). However, in many cases and in all cultures, exceptionality was associated with sinfulness. It was looked upon as a kind of curse or punishment, either because of a lack of faith or because of some failure in the individual or in his or her parents. This mistaken attitude has fundamentally colored society's perception of differences, and even today the attitude has not entirely been dispelled.

Since earliest recorded history there have been instances of humane treatment of those who were different or disabled. In Athens, around 600 BC, a famous lawgiver named Solon designed a system for providing care for soldiers who had been disabled in war. However, Solon's attitude was uncommonly enlightened. In the Middle Ages, the lives of exceptional people were often full of suffering, and customarily they benefited from little or no support from society. Both Martin Luther and John Calvin, important Christian reformers in the sixteenth century, stated their belief that disabled individuals had no souls and that therefore society had no responsibility to them at all. Those who were deaf, mentally retarded, epileptic, and blind, as well as some especially gifted or talented individuals, were often thought to be possessed by demons and sometimes were subjected to exorcisms to purge the "devils" from them.

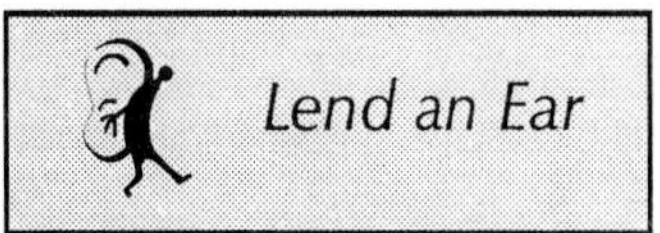

"In Germany, the Nazis first came for the communists, and I did not speak up, because I was not a Communist. Then they came for the Jews, and I did not speak up, because I was not a Jew. Then they came for the trade unionists, and I did not speak up, because I was not a trade unionist. Then they came for the Catholics, and I did not speak up, because I was not a Catholic. Then they came for me... and by that time, there was no one to speak up for anyone."
—Pastor Martin Niemoeller, German Evangelical Church

Figure 1.5
All the world's major religions teach against prejudice in favor of tolerance, unity, and love.

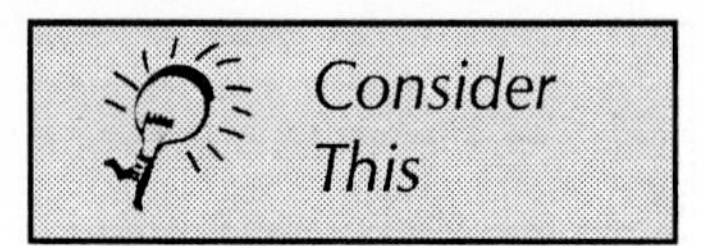

Why must we continually redefine what constitutes human difference?

The 18th and 19th centuries brought the beginning of efforts to educate and to care for exceptional people. There was more treatment for people who were mentally ill and mentally retarded, and programs were instituted to educate people who were deaf, mute, and blind. Special schools for special conditions—such as asylums for people who were blind—began to be established in the United States as well as in Europe. Programs within public schools in the United States were begun in the early 1900's, segregating exceptional children into special classes apart from the main student body. Since that time, programs for students with special needs have steadily grown and developed, with an emphasis on inclusion rather than exclusion.

While there have been flashes of enlightened attitudes and actions toward human diversity throughout history, in general our treatment of individuals with differences has been shameful. In every period of history so far, and in virtually all cultures, there has been a stigma attached to difference. Societies have neglected, persecuted, and sometimes exterminated those of its members who exhibited noticeable variation from the perceived norm. Since perception is always subject to change, and all of us fall into some category of exceptionality, it behooves us to continue to grow in humanity and in understanding. The best way to do that is to educate ourselves (see Figure 1.6). Intolerance is the direct result of ignorance.

Anthropology
Art History
Communications
Comparative Religion
Demography
Ethics
Genealogy
Geography
History
Language
Linguistics
Literature
Philosophy
Political Science
Psychology
Sociology
Special Education

Figure 1.6
A variety of fields contribute to our understanding and appreciation of human diversity.

The on-going enactment of human rights legislation and the increasing awareness of the contributions of diverse people to our society gives hope that we may be entering a new century in which differences are not only recognized and tolerated but appreciated and honored. In the past two decades there has been phenomenal progress in the fields of medicine and technology. We must continually redefine exceptionality due to increased understanding of the causes of some categories of diversity, medical advances in the cure and treatment of some conditions, and new technology which enables people to function despite previously limiting handicaps. It is an exciting time to be studying the subject of human diversity.

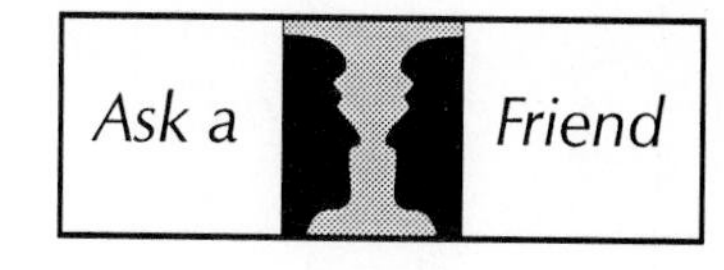

Has a label ever hurt your self-esteem or stigmatized you? Ask a friend about his or her experiences with labels.

Terms and Labels

The terms you use to describe or identify individuals who are different should be positive, current, and correct. The goal is not to expand the lexicon of political correctness. But it is important to think carefully about the words you choose to use. You can probably think of many inappropriate terms which have been used to identify people with diverse characteristics. The acceptability of terms changes with time, and a word may be the term of choice of one group or individual and be offensive to another. For example, some people who are hearing impaired prefer to be described as "hearing impaired," yet the National Theater For The Deaf uses the term "deaf." The best way to be sure that you are not giving offense is to go to the expert. Ask a friend or acquaintance who is diverse what is most acceptable to them (see Figure 1.7).

Labels serve an important purpose. They are the way we attempt to identify and describe people. They allow us to distinguish characteristics which may facilitate communication among professionals. For example, psychologists and educators may use labels to discuss the special needs of an individual or of a group of people. On the other hand, labeling may stigmatize a person, contribute to low self-esteem, and cause him or her to suffer discrimination. Once labels are applied to someone, particularly by an official evaluator such as a doctor, teacher, or psychologist, it may be difficult to remove the label. Consequently, we should carefully give consideration to the accuracy of those labels and to their potential effects.

Figure 1.7
Before you stamp someone with a label, ask what terminology is acceptable.

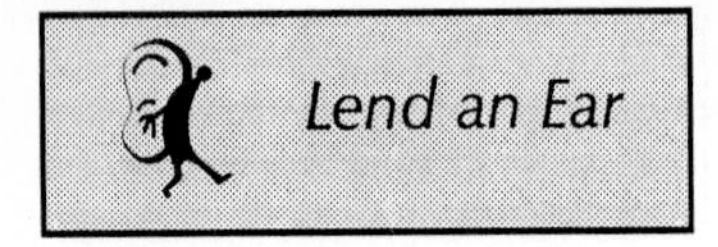

"We are of course a nation of differences. Those differences don't make us weak. They're the source of our strength." —Jimmy Carter, president

In any culture, some labels are generally thought to be positive and others negative. Most of us would have a positive reaction to the label "intelligent," but we might have very different ideas what we mean by that label. Some of us say our dogs are intelligent if they can roll over, for instance. In our culture, "overweight" is usually a negative label, but a fashion model might be called overweight even though, by the standards of the society, she is thin. Words have different connotations, or meanings, for different groups, and the effect of labeling, as we have seen, can differ from person to person. The best guideline is to always remember that the person comes first and the label simply identifies a specific characteristic about that person. For instance, don't say "Moslems practice beautiful traditions," but rather "People of the Muslim faith practice beautiful traditions."

Attitudes

Consider your attitude toward people who are different. When you meet someone who is using a wheelchair or someone of a different race, how do you react? How do you feel? What about someone who speaks with a heavy accent or who uses gestures and sign language for communication? Or a street person? Or two men or two women dancing together? Are you uncomfortable? Do you try to avoid the person? The attitude that others display toward people who are exceptional is critical.

Attitudes are learned early in life when children interact with others and observe the human interactions of family members and friends. The child who hears his or her parents make negative remarks about diverse people, or who observes playmates ridicule and tease other children who are different, will probably be greatly influenced by these experiences. The child who grows up in an environment where individual differences are noted and treated with respect will more likely not have negative expectations, feelings of discomfort, or negative attitudes.

Can you improve the attitudes of children and adults toward diverse individuals? You certainly can by being a good model and following these guidelines.

Always display your respect and comfort. It is important for you to set a good example. If you interact with everyone in a respectful manner and if it appears that you are comfortable in your interactions, your friends, family members, fellow employees, and your children will follow your lead. Remember that others will copy your behaviors.

Don't worry about what other people may or may not be thinking. If you are heterosexual and are talking to someone who is openly gay, or if you are a Christian talking to a Hindu, don't worry what other people may or may not be thinking about you. Communicating with diverse people in no way compromises your identity. If other people react negatively to your association with someone who is different, remind them that you were talking to the person, not to the label. If you are seen talking to an individual who is different, the only thing it says about you is that you are open, friendly, and non-judgmental.

Stop someone who is ridiculing or joking about people who are diverse. If a person begins to ridicule or joke about someone, challenge that person to stop. If you sit there and laugh along with the crowd, you are just as guilty as the joke teller. By participating you are giving others the impression that you condone such negative behavior. It really takes guts to stop a friend or co-worker, but if you do so in a serious and positive manner, others will no doubt respect you for having the courage of your convictions. Comments such as, "Excuse me, but the racist joke you are telling is making

Now is the time to get rid of all the negative terms you can think of for describing exceptional people. You've probably heard many. List 20 of those terms. Then cross them out as a symbol of your decision never to use them again.

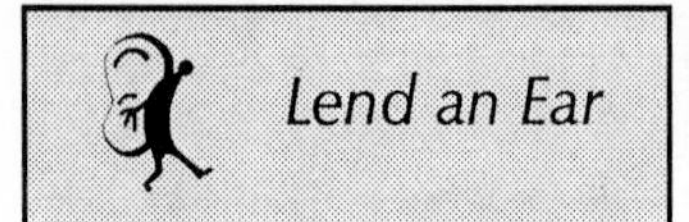

"My father taught me how to understand and be sensitive to others, which is probably the most critical aspect of leadership." —David Morehouse, former Army airborne ranger commander

me uncomfortable," or "If you don't mind I'd appreciate your saving any jokes about male-bashing for a time when I'm not around" will usually work. Another way to handle the situation would be to tell the group you are leaving because of the stereotypical jokes which are being told. Then walk away.

Tell children about differences. Encourage children to ask questions as their curiosity is quite normal. If children are given correct information in a matter of fact manner, they will perceive that being different is not mysterious or something to fear. People tend to fear that which is unknown, so it is critically important to teach children about human diversity so that all possible fears about exceptionalities are eliminated.

Expect normal behaviors and achievement from diverse individuals. If you think that a non-native speaker of English or an individual with a hearing impairment will be unable to get the job he or she is applying for, you will be hampering the employment success of that person. If you expect that a friend with mental retardation can never learn to ride a subway, you are setting a major roadblock in the way of that person's learning to use public transportation. The expectations of influential people, such as parents, teachers, and friends, have a strong effect on people's confidence level and motivation to achieve. The potential of people who are diverse is usually only limited by the opportunities to learn and the inappropriately low expectations of significant others.

Be yourself around those who stand out as different. If you are normally crabby, be crabby. If you are normally friendly, be friendly. Putting on an act around people who are different will be recognized as patronizing or demeaning behavior. It is not necessary for you to sprint ahead and open a door for someone in a wheelchair unless you usually do that for everyone. If you are taking a leisurely walk and normally say hello to people you meet, then by all means say hello to that person with a guide dog who is walking by. If you usually ignore people whom you don't know, and you don't know that person with an oxygen tank, then ignore him or her as well. Every person wants to be treated like everyone else, and they can tell when you are acting. Therefore, be yourself and don't change your behaviors around those who are different.

Use your common, everyday vocabulary. The vocabulary you commonly use should be fine with anyone you come across. With people who are blind it is appropriate to say, "It's nice to see you," or "Isn't it a beautiful day?" It is able acceptable to say to a friend of yours who uses a wheelchair, "Let's take a walk to the park." People who are different use common, everyday language all the time, so don't feel the least bit awkward using your everyday vocabulary. If you stop to think about the "right" words to use, you will come across as uncomfortable and the substitute words you select will probably be wrong.

Suggest activities in which you are interested. Your fellow human beings may enjoy the same hobbies as you do. If you like tennis, ask your friend who has epilepsy to play. If you are going to a theme park, invite your friend who is a Buddhist to accompany you. If you want to go to an art museum, don't hesitate to invite your friend who is visually impaired along. Diverse people should not be excluded from activities just because *you* think those activities would not be appropriate for them. Involve your friends in your life. They are fully capable of selecting the activities which they will enjoy and which fit their interests (see Figure 1.8).

Figure 1.8
Diverse individuals may enjoy the same activities as you.

Look at but never stare at someone who is different. It is fine to look, note any differences in your mind, and then go on with your interaction with an exceptional person in a normal fashion. Staring at someone who is dressed differently, or disfigured, or short statured is rude and will make that person feel uncomfortable. Some exceptional people will stare right back or will bluntly ask you why you are staring. It is very appropriate, however, to look in a normal fashion at people who appear different with whom you are interacting. Avoiding eye contact, or looking away, is just as rude as staring.

Talk to the person who is exceptional and not just to his or her companion. When interacting with an exceptional person who is with a family member or companion, be sure to address the person rather than the family member or companion (see Figure 1.9). A person from another country can usually answer questions or respond appropriately in conversation without assistance. Likewise, if you are with a friend who has a disability, don't respond for the friend. When put in that situation, such as a waiter asking you what your exceptional companion would like to eat, don't respond for your friend. Politely suggest to the waiter that he or she should ask your friend. Although that situation may be a bit uncomfortable, it will be a good experience for that insensitive restaurant employee.

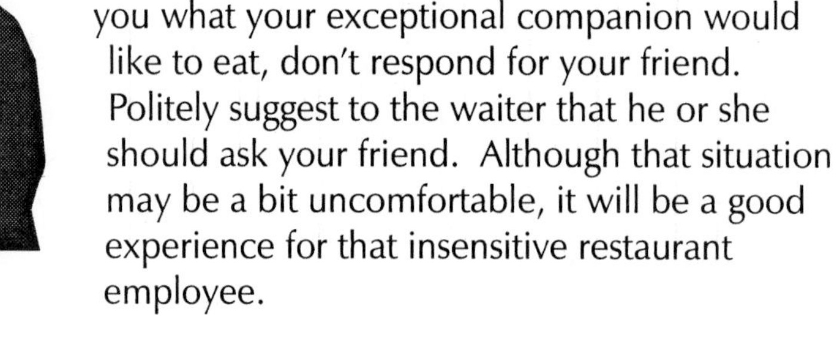

Figure 1.9
Talk to the person who is exceptional and not just to his or her companion.

The guidelines suggested above should assist you in your efforts to improve your interactions with exceptional people, and they should serve as excellent examples for children and other adults who observe your behaviors. By interacting correctly and comfortably, you will greatly aid those people who are uncomfortable or afraid due to ignorance, myths, and misconceptions.

Conclusion

In this first chapter, you have had an opportunity to begin your exploration of the subject of people who are diverse. You have been introduced to some of the labels which have been applied to individuals who are considered different and to the effects those labels can have. You have learned correct terms, along with some guidelines for appropriate interactions, and you should have begun to consider the importance of positive attitudes. You can see that each of us, every day, falls into a category of being different at some time. As you learn more about how differences affect individuals, you will enjoy enhanced self-awareness and a keener sense of those around you.

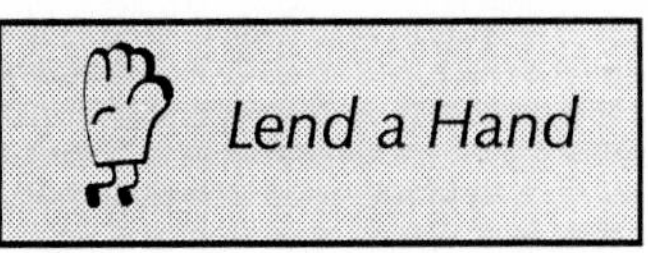

At the neighborhood and city levels, each of us can foster community involvement, non-violence, and compassion. By working for change in our neighborhoods, jobs, families, circles of friends, and within ourselves, we can bring about progress on a global level.

References

U.S. Department of Education. (1994). *Schools in the United States: A statistical profile.* Washington, DC: Office of Educational Research and Improvement, National Center for Educational Statistics.

Suggested Readings

Angelou, M. (1989). *I know why the caged-bird sings.* New York: Literacy Volunteers of New York City.

This is the autobiography of an African-American woman, telling how she overcame the social biases of growing up poor and black to become a fully realized and successful person.

Brown, C. (1976). *The children of ham.* New York: Stein and Day.

This book offers personal accounts of African-American youth in the slums of Harlem and their proposed solution to their social ills.

Frank, A. (1956). *The diary of Anne Frank.* New York: Random House.
A young Jewish girl's spirit triumphs in the face of Nazi occupation, just before she and her family are exterminated on the basis of their race and religion.

Momaday, N. S. (1990). *The way to rainy mountain.* Albuquerque: University of New Mexico Press.
This is an account of the legends of the Kiowa Indians and how their ancient religion and culture were suddenly and brutally eradicated.

Monette, P. (1992). *Becoming a man: Half a life story.* New York: Harcourt Brace Jovanovich.
Provided in this autobiography are a gay man's poignant memories of growing up, coming out to his parents, and battling AIDS.

Neusner, J. (1994). *World religions in America.* Louisville, KY: John Knox.
This book examines the faiths of African-Americans, Hispanics, Native Americans, and all other major denominations. It also examines the subjects of women and religion, politics and religion, and society and religion.

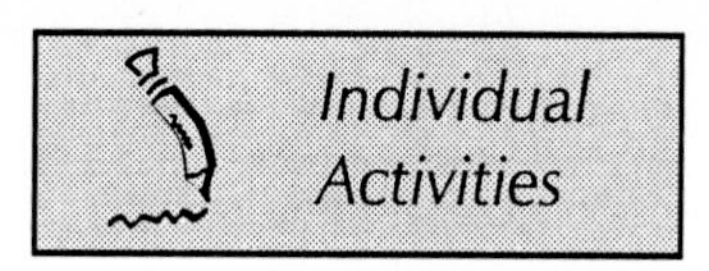

1. You are an exceptional person! Every single person on earth has a unique talent. There are things you do much better than others, and there are some things you don't do as well. List eight things that make you a person who is different. Circle those which are extremely higher or lower than what you consider normal.

________________________ ________________________

________________________ ________________________

________________________ ________________________

________________________ ________________________

2. When you were a child, didn't you hate being called names? Were you ever called a name or did you have a nickname that stuck with you for a while? What effects did that label have on you? On others? Relate your discussion to people who are diverse.

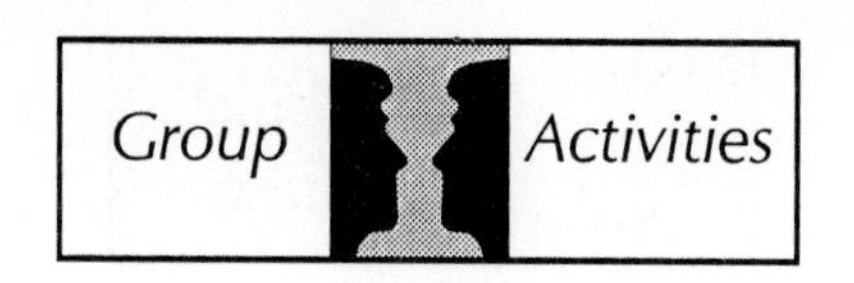

1. Get into a group of 4 or 5 people—people whom you don't know well. Have each person write down a list of 5 adjectives which are guesses about the skills and weaknesses of each person in the group. Now discuss what everyone in the group wrote. Were the people in your group right or wrong about you? Did you find out something about yourself you didn't already know? How correct were you about others? Welcome to the world of labels!

Reactions:

2. Review the guidelines presented in this chapter for interacting with diverse people. Now, as a group, discuss and rank these guidelines from most to least important.

Reactions:

Reflection Paper 1.1

In the quotation that opened this chapter, Helen Keller suggests that intolerance has destroyed more lives than the most hostile forces of nature combined. Do you agree or disagree with her statement? Why?

Reflection Paper 1.2

Throughout history, society continually changes the way it defines differences and exceptionalities. Discuss why you think that might be so.

Notes

Chapter 2 The Culture of Human Diversity

Chapter Sections

- Introduction
- Aspects of Culture
- Cultural Distinctions
- Rethinking Culture
- Key Cultural Concepts
- Interactions
- Conclusion
- Individual and Group Activities

"Culture is not life in its entirety, but just the moment of security, strength, and clarity."
—José Ortega y Gasset, philosopher.

Objectives

- Explain how cultural identity is determined.
- Explain how cultural identity is passed along from one generation to the next.
- Distinguish between "ideal" and "real" culture.
- Distinguish between "explicit" and "implicit" culture.

Introduction

The term *culture* is frequently misused. *Culture* is not a synonym for *civilization*. Think of civilization as a universal quality inherent in all people. In other words, we all have the ability to behave in a civilized manner, exhibiting compassion and practicing cooperation. Culture, on the other hand, is a subjective—rather than universal—human quality. Our culture is made up of the ideas, customs, skills, arts, interests, and emotions of our people (see Figure 2.1). We communicate and pass along these ideas and customs to succeeding generations, and thus our culture survives and prospers. Civilization is a "social climate in which people in differing groups can deal with each other in ways that respect cultural differences" (Cleveland, 1994). Diversity strengthens our species and adds threads to the tapestry of our lives. It exponentially increases the possibilities for progress and positive change in the world because rather than seeing from only one point of view, it allows us to see from many. We would be sadly diminished if the music of the universe consisted of only one note. We would be even more sadly diminished if human beings were free of differences. Too often we speak of tolerating difference when

Figure 2.1
Members of a culture share many important qualities. Diversity is the pot that holds culture together. Cultures are always changing and growing. Understanding other cultures enriches and strengthens your own culture.

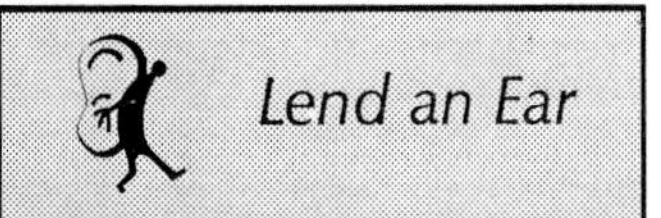

"The knowledge of another culture sharpens our ability to appreciate our own."
—Margaret Mead, anthropologist

we should always be respecting and celebrating it.

Sometimes science fiction films and books portray a world where diversity has been limited or eradicated. Without exception, those images are frightening in their sameness and colorlessness. We recoil from the prospect because we instinctively know that to destroy difference is to destroy our own individuality–the very thing which makes us feel human–our uniqueness. The lesson we must learn and never forget is that only by respecting and honoring the difference in others can we preserve our own. The English poet John Donne wrote, "Every man's death diminishes me." The corollary to that truth is that everyone's difference enhances me.

Aspects of Culture

Our world is like an intricate jigsaw puzzle, every piece representing a unique culture. It is virtually impossible to escape the influence of culture, no matter what you do or where you go. When rivers meet, they flow, but cultures often just collide. In the calming of the collision, we can all be heard more clearly stating who we are (Center for Cultural Interchange, 1996). Inter-cultural communication skills are a necessity for everyone, not just for the culturally "deprived" or distinct, but for all people as cultural beings.

All cultures have a history and a heritage which is carefully handed down. Cultures are identifiable because they share certain characteristics. People within the group have ways of telling who is and who is not a member. Members share a language through which cultural interactions take place, certain group values are shared, and various social and behavioral patterns are followed. Members follow special rules of etiquette when communicating with others, introducing themselves to strangers, getting someone's attention, and leave-taking, for example (see Figure 2.2).

Figure 2.2
Members of a culture share special rules of etiquette and other social customs.

Members of each culture also have their own unique slang expressions and figures of speech. You can learn more about these codes of behavior and expression by asking what is appropriate in certain situations. However, some of these rules may not be meant for general public knowledge. Such cultural interactions are often kept private so that members of the specific community may identify outsiders easily (Bienvenu & Colonomos, 1993).

What are some slang expressions typical of your culture?

Cultural Distinctions

In anthropology, important distinctions are made between "ideal" and "real" culture (Arvizu, Snyder, & Espinosa, 1980). Ideal culture refers to what people say they believe or how they think they should behave. Aspects of ideal culture are frequently expressed in proverbs, stories, myths, and jokes. These may contrast sharply with the real culture, which is how individuals actually behave in specific situations.

Some elements of culture operate at a conscious level of awareness, whereas others do not. Implicit (covert) culture includes elements hidden or taken for granted to the extent that they are not easily observable or consciously recognized by individuals. Attitudes, fears, values, religious beliefs, and assumptions are common elements of implicit culture. Explicit (overt) culture, on the other hand, is visible and can be described verbally. This includes speech, tools, styles of dress, and concrete behaviors (see Figure 2.3).

Figure 2.3
Explicit culture includes dress and other visible elements.

Having made these distinctions, it must be noted that the scientific theories of culture do not always pertain to the realities of people who live together. When we examine particular cultures up close, we find that the individuals do not *necessarily* share beliefs and ways of life. As social scientist Amelie Rorty notes, a group of people may be held together by the joys of

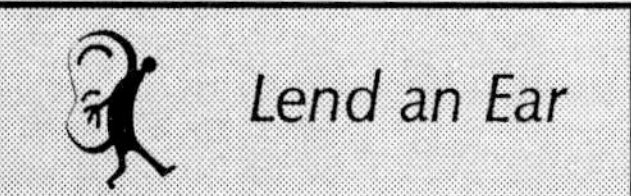

"The choice of a point of view is the initial act of a culture."
—José Ortega y Gasset, philosopher

arguing with one another, even about the most fundamental issues! "How finely, at what level of generality do a people share customs, dress, music, and so on?" Rorty asks. "In many cultures, men and women, the young and the elderly, priests and merchants, the high and low born do not have the same practices or concepts. They may recognize—and be committed to—their practical, economic interdependence, but they often sincerely profess not to understand one another; and they often identify themselves by their internal contrasts, rather than by their alleged presumptively shared culture" (Rorty, 1995, p. 161). Many people are quick to identify themselves with a particular culture, but we shouldn't lose sight of the how much *diversity* exists within even the smallest of groups.

Rethinking Culture

According to Harland Cleveland, president of the World Academy of Art and Science, no single culture is or can be complete in itself. Cultures keep redefining themselves, he explains, by mixing and matching with other cultures, "not only through getting to know people who look, act, and believe differently but through exposure in a more open electronic world to new faiths and fashions, new lifestyles, workways, technologies, clothing, and cuisines." Every culture, then, is in a constant state of flux. "But many millions of people in this time of uncertainty and insecurity believe that their best haven of certainty and security is a group based on ethnic similarity, common faith, economic interest, or political likemindedness," Cleveland says. "The fear that drives people to cleave to such 'primordial loyalties' makes it harder for them to learn behaviors consistent with tolerance of others who may be guided by different faiths and loyalties" (1994, p. 756).

The only way to fight such fear is to avoid uncritical, narrow-minded thinking. In the words of

Danny Weil, an expert on multiculturalism and critical thinking, the way to begin to think fair-mindedly and critically is to exercise reciprocity. Reciprocity simply means "to imaginatively place oneself in the 'shoes' of others often diverse in thought from ourselves, to consider strengths and weaknesses of opposing cultural and political points of view, and to overcome our sometimes ego-centric tendencies to wed ourselves blindly and uncritically to one belief or another without the benefit of self-examination and critical analysis" (Weil, 1993, p. 211).

The first step to overcoming egocentrism is to ask yourself these key questions (Derived from Weil, 1993):

- How do I arrive at my conclusions and choices?
- Upon what assumptions do I base my inferences?
- What evidence do I have to support my beliefs?
- What other points of view inform the bank of data and evidence I use to support my assumptions and consequent decisions and actions?

Uncritical thinking, Weil explains, is inherited from secondary sources, such as television, movies, parents, friends, institutions, and teachers. "The uncritical mind looks for stereotypes and simplistic categories in which to conveniently place people, things, and places. . . . Without the benefit of critical reasoning within and about diverse cultural points of view, the human mind becomes at peace with internalized cultural stereotypes, falsehoods, prejudices, and biases, and can accomplish little to help transform the world lived in and with others." In other words, the uncritical mind cannot fully participate in or take advantage of another culture, much less its own (see Figure 2.4). If we are to succeed in living and thriving together as human beings with diverse cultures, "we must become actively engaged in dialogue about diversity, with an interest in developing fair-minded reasoning in the search for personal, social, and political transformation" (Weil, 1993, p. 211).

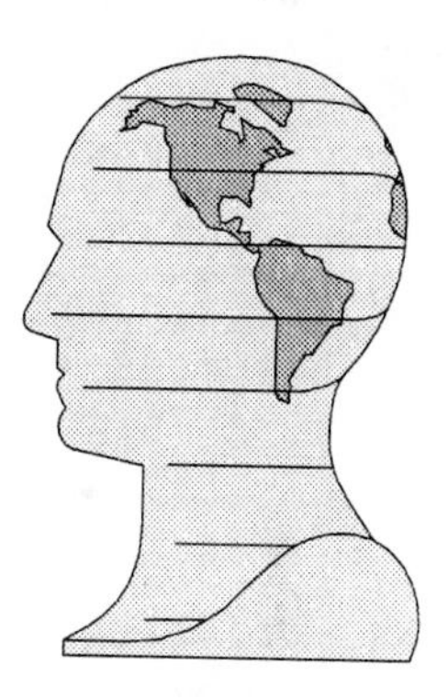

Figure 2.4
Only a mind which is actively aware of diverse cultural viewpoints can make positive change in the world.

Ask a friend, either an immigrant or not, to share how he or she went through a process of acculturation.

Key Cultural Concepts

The following concepts need to be considered when thinking and talking about culture and diversity:

Acculturation. This is the process by which the members of a society are taught the elements of that society's culture. Everyone goes through an acculturation process, immigrants and non-immigrants alike.

Common Culture. This refers to the common body of knowledge that allows diverse people to communicate, to work together, and to live together (Hirsch, 1993).

Counter Culture. This refers to a protest movement in thelate 1960s, in which young people formed their own culture in opposition of the culture of Middle America. This movement was epitomized by "Hippies" and the Woodstock festival.

Cultural Encapsulation. Similar to *ethnocentrism,* this is a closing-off of one culture from others.

Cultural Literacy. This refers to the names, phrases, events, and other items that are familiar to most literate persons in a given culture (Hirsch, 1993). To be culturally literate is to know the shared information that binds your culture together.

Cultural Pluralism. This refers to the belief that diverse groups coexist within American society and maintain a culturally distinct identity (see Figure 2.5).

Figure 2.5
Through cultural pluralism, diverse groups coexist within American society.

Cultural Understanding. This refers to the process of learning about other cultures in order to foster social harmony and growth.

Ethnocentrism. This is the belief that one's cultural ways are not only valid and superior to other people's, but also universally applicable in evaluating and judging human behavior.

Heritage. This refers to something that belongs to a person by reason of his or her birth. A person may be born into one culture and later ignore or reject that cultural inheritance in favor of another.

Intercultural. This means "between" or "among" cultures. Intercultural interactions involve mutual or reciprocal exchanges.

Melting Pot. This refers to the concept that many cultures can blend into one. This concept has historically been a component of American culture. However, it no longer is widely accepted as a goal within our society.

Microculture. This implies a greater linkage with the larger culture, and emphasis is often put on the degree to which the microculture acts to interpret, express, and/or mediate the ideas, values, and institutions of the political community.

Multicultural. This refers to a number of diverse traditions, customs, arts, languages, values, and beliefs existing side-by-side.

Subculture. This is a term used frequently by sociologists to refer to a social group that shares characteristics that distinguish it in some way from the larger political society (usually called macro-culture) of which it is a part.

Youth Culture. Leisure, lifestyles, clothing styles, musical tastes, and peer-group values typically define youth

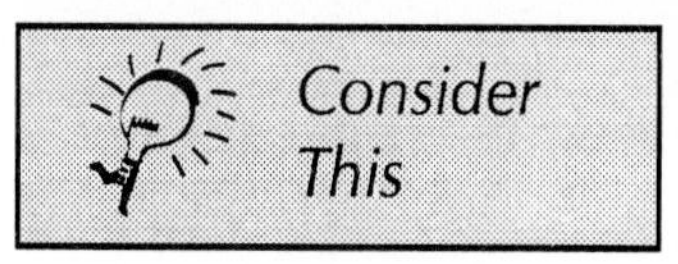

When you speak, use the pronouns "you" and "your" rather than "us," "we," and "our." "We" might suggest "everyone but you" to the person you're conversing with, while the personal "you" signals that you are speaking to that individual.

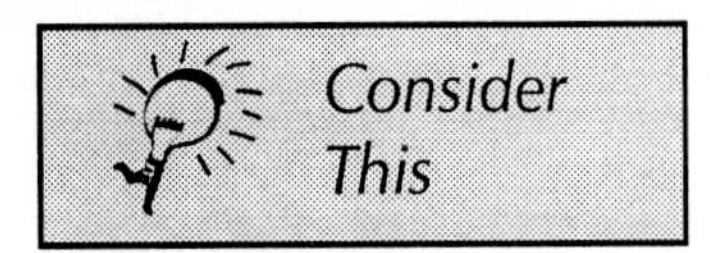

As you encounter people in your daily life, briefly look them in the eye. This helps them recognize that you see and value them.

cultures. Youth cultures exhibit particular rules of behavior which are often considered as deviant or in opposition to the dominant group.

Interactions

Pull out a dollar bill. There is a slogan on our currency, written in Latin: "E pluribus unum." It means "From many, one," or "Wholeness incorporating diversity" (see Figure 2.6). Look at this phrase every time you exchange money with someone, whether it be a friend, a store clerk, a bank teller, or a starving person. If you find yourself feeling uncomfortable due to someone's cultural differences, give a dollar bill to that person and explain the meaning of the Latin slogan.

Figure 2.6
The dollar bill features the slogan "E pluribus unum," meaning that diverse individuals are unified.

Cheerfully acknowledge differences. Equality is not the product of similarity. Rather, it is the cheerful acknowledgment of difference (Cleveland, 1994). We are all equal in the sense that we are all equally *different!*

Identify the common culture. When you meet someone of a different culture, try to identify the shared knowledge that allows the two of you to communicate, to work or play together, and to live together.

Identify differences within your own culture. Within every cultural sub-group is a vast array of diversity. Before you quickly lump yourself or someone else into a cultural category, remember the number of differences which exist in every group. It is differences which foster growth, add complexity and richness, and make a culture dynamic.

Stop stereotypical jokes. When friends or relatives tell stereotypical jokes, they are exhibiting uncritical, narrow-minded thinking. Explain why you find that brand of humor offensive. Be a good example yourself by never lumping individual people into broad categories or making fun of others based upon stereotypes.

Live by the "Creed for Citizens of a Diverse World." Griessman (1993) provides insight to appropriate interactions and ideas for respecting culture:

> I believe that diversity is a part of the natural order of things-as natural as the trillion shapes and shades of the flowers of spring or the leaves of autumn. I believe that diversity brings new solutions to an ever-changing environment, and that sameness is not only uninteresting but limiting. To deny diversity is to deny life- with all its richness and manifold opportunities. Thus I affirm my citizenship in a world of diversity, and with it the responsibility to:
> **Be tolerant.** *Live and let live. Understand that those who cause no harm should not be feared, ridiculed, or harmed-even if they are different. Look for the best in others.*
> **Be just** *in my dealing with poor and rich, weak and strong, and whenever possible, to defend the young, the old, the frail, the defenseless. Avoid needless conflicts and diversions, but always be willing to change for the better that which can be changed.*
> **Seek** *knowledge in order to know what can be changed, as well as, what cannot be changed. Forge alliances with others who love liberty and justice.*
> **Be kind,** *remembering how fragile the human spirit is.*
> **Be generous** *in thought, word, and purse. Live the examined life, subjecting my motives and actions to the scrutiny of mind and heart so to rise above prejudice and hatred.*
> **Care.**

Issues Culturally Diverse Individuals May Face

- Assimilation v. maintaining a cultural identity
- Being labeled as a foreigner or minority
- Perceived as stealing jobs
- Homesickness
- Finding peer support
- Language issues
- Perceived as illegal alien
- Cultural differences
- Overcoming stereotypes
- Separation from family
- Inter-cultural relationships
- Celebrating of heritage
- Family origins
- Self-esteem and self-image issues
- Recognizing diversity within their community
- Feeling out of place
- Equal access or opportunity
- Prejudice or racism

Untrue Stereotypes of People Not of the Majority

- Have had a disadvantaged background
- Need or want exceptions to get into school or a particular job
- Do not understand English
- Are in the U.S. illegally
- Are experts on their culture
- Spend all their money in their home country
- Have a chip on their shoulder
- Want special rights
- Have a large family

Conclusion

Diversity enriches the lives of individuals and strengthens society as a whole. Diversity expands the possibilities for progress, evolution, and positive change in the world because it allows us to see from many different perspectives. If human beings were free of differences, the quality of our lives would be profoundly diminished. It is time to stop merely tolerating difference and start celebrating it.

References

Arvizu, S. R., Snyder, W. A., & Espinosa, P. T. (1980, June). Demystifying the concept of culture: Theoretical and conceptual tools. *Bilingual Education Paper Series, 3*(11). Los Angeles: Evaluation, Dissemination and Assessment Center, California State University, Los Angeles.

Bienvenu, M. J. & Colonomos, B. (1993). *An introduction to deaf culture: Rules of social interaction.* Burtonsville, MD: Sign Media.

Center for Cultural Interchange. (1996). *Host family handbook.* St. Charles, IL: Author.

Cleveland, H. (1994). The limits of cultural diversity. *Vital Speeches, 60,* 756.

Greissman, B. E. (1992). *Diversity challenges and opportunities.* New York: Harper Collins.

Hirsch, E. D. (1993). *The dictionary of cultural literacy.* Boston: Houghton Mifflin Co.

Rorty, A. O. (1995). Rights: Educational, not cultural. *Social Research, 62,* 161.

Weil, D. (1993). Towards a critical multicultural literacy: Advancing an education for liberation. *Roeper Review, 15,* 211.

Suggested Readings

Duval, L. (1994). Respecting our differences: A guide to getting along in a changing world. Minneapolis, MN: Free Spirit.

Discussions of accepting, respecting, and celebrating differences are provided in this guide.

Gold, S. & Kibria, N. (1993). Vietnamese refugees and blocked mobility. *Asian and Pacific Migration Journal, 2(*1), 27-56.

This journal examines data from published sources and ethnographic studies of the economic situation of Vietnamese refugees in the United States.

Kroeber, A. J. & Kluckhohn, C. (eds.). (1954). *Culture: A critical review of concepts and definitions.* New York: Random House.

This is an invaluable overview of cultural concepts and definitions.

Mindel, C. H. & Habenstein, R. W. (Eds.). (1988). Ethnic families in America: Patterns and variation (3rd ed.). New York: Elsevier Publications.

This book examines ethnic family strengths and needs as well as historical background and demographic characteristics.

Vedder, R & Gallaway, L. (1993). Declining black employment. *Society, 30(*5), 57-63.

This article explores income inequality during declining African-American employment, examines current welfare systems, and suggests ways to improve the economic disadvantages of minority groups.

Notes

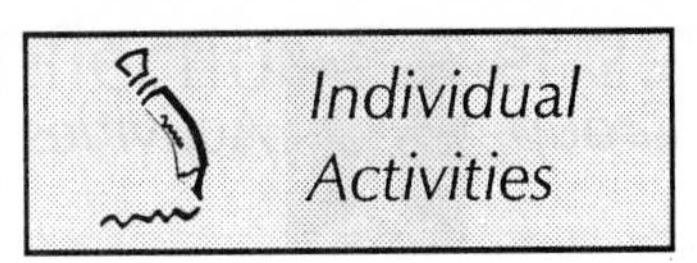

1. The native or mother tongue is a key element of culture. This activity encourages you to examine the particular dialect you speak. If you are a native speaker of American English, look in a dictionary printed in England and find examples of words or phrases that are unfamiliar. For example, Americans say "elevator," whereas the British say "lift." Write down six words or expressions that are common to your particular dialect of English. (If you are a native speaker of another language, do the same exercise with an appropriate dictionary.)

______________________ ______________________

______________________ ______________________

______________________ ______________________

2. In the space below, write a creed for yourself to live by as a citizen of the world.

Reflection Paper 2.2

Discuss the difference between ideal and real culture. Give examples in your discussion.

Notes

Chapter 3 The Rights of Diverse People

Chapter Sections

- Introduction
- The Laws of Human Diversity
- Bringing About Change
- Conclusion
- Individual and Group Activities

"The great law of culture is: Let each become all that he was created capable of being; expand, if possible, to his full growth; resisting all impediments . . . and show himself at length in how own shape and stature, be these what they may."
—Thomas Carlyle, scholar

Objectives

- Explain the rights guaranteed by the Equal Protection of Laws.
- Identify the major laws which ensure opportunities for diverse people.
- Describe legal issues diverse people may face.

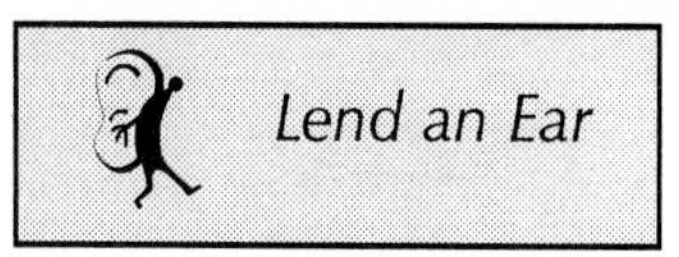

"There is but one law for all, namely, that law which governs all law, the law of our Creator, the law of humanity, justice, equity—the law of nature, and of nations."
—Edmund Burke, statesman and author

Introduction

Although we have seen that exceptionalities fall into many categories, the social climate in a culture tends to affect many of those persons at once. In other words, interest in promoting the rights of all people in general leads to attention to the rights of specific groups. In the United States, the labor movements of the 1930s helped to focus some attention on physical impediments to work and on the needs and rights of children. The 1950s and 60s saw the beginning of the civil rights movement, which focused on the need for social change in regard to African Americans. Gradually, the methods which proved to be effective in raising the consciousness of the American people about the plight of African Americans, such as education and free speech, came to be implemented in the cause of women's rights, gay rights, disability rights, and the rights of all individuals to freedom from discrimination. In all categories of exceptionality, people deserve and demand freedom from personal, social, and economic discrimination (see Figure 3.1).

While we work for fundamental change in individual attitudes which will promote understanding and tolerance of difference, it is important to protect human and civil rights through

Figure 3.1
Justice is said to be blind, guaranteeing every person equal rights regardless of superficial differences.

Why is it important to protect human and civil rights through legislation?

Legal Issues Diverse People May Face

- Dealing with the police
- Knowing when a lawyer is necessary
- Finding the right lawyer
- How to conduct oneself during an arrest
- Avoiding legal problems
- Knowing one's legal rights
- Understanding the language of the law
- Understanding how the law works
- How to report a violation of one's rights
- Knowing what constitutes a crime
- Legal expenses
- Becoming a crime victim
- Knowing the rights of the accused
- Witnessing a crime
- Going to court
- Courtroom protocol
- Trial procedures
- Serving on a jury
- Being a trial witness

legislation. Local, state, and federal laws insure opportunities for employment, provide for appropriate education and medical care, and seek to secure access to public buildings and streets for everyone. The struggle for equal opportunity has not been an easy one, nor is it over. However, thanks to the many court decisions which have upheld individual rights, state government and the federal government have gone far to establish a public policy through legislation which recognizes the right to be different.

The following is a brief discussion of some major areas of legislative protection. It is important for every individual in this society to be aware of the laws protecting our rights and the rights of others. Even a cursory study of this list will help you to see some of the difficulties people with exceptionalities may encounter. Such a study should raise your consciousness and increase your awareness of the ways in which individuals are penalized for being different in schools, in the workplace, and in social situations every day.

The Laws of Human Diversity

The following overview examines the legal aspects of various categories of human diversity. Because state and local laws change frequently, this examination will focus primarily on selected federal legislation that affects all Americans. It is important to recognize that this information is not all-inclusive. Additional reading and study are needed to achieve a complete understanding of laws which impact persons who are diverse.

Age. As longevity increases, our society has the benefit of more and more older citizens who are increasingly physically able to perform in the workplace. These older Americans offer wisdom and experience from which we can all profit. Due to the changing nature of work in the information age, it is even more clear that there is no justification for discrimination on the basis of age.

The Older Americans Act of 1965 (OAA) is designed to safeguard the rights of senior citizens. The Act focuses on health care, Social Security, and employment. The Administration on Aging (a division of the Office of Human Development Services) is the principal agency designated to carry out the provisions of the OAA. Senior advocacy groups in Washington monitor policy on Social Security, Medicare, and Medicaid.

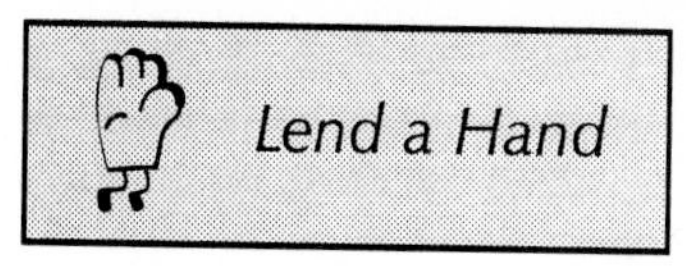

It takes individuals:

- to demand freedom.
- to raise consciousness.
- to enact legislation.
- to carry out that legislation.

Disability. The Rehabilitation Act of 1973 provides for a "Barrier-Free Environment" for people with disabilities (see Figure 3.2). The Act defines these individuals as persons who have a physical or mental impairment which substantially limits one or more major activities, have a record of such an impairment, or are regarded as having such an impairment. The Act requires any program or activity receiving federal monies to provide equal access and opportunities for people with disabilities. This includes people with cancer, heart disease, diabetes; cerebral palsy, epilepsy, mental illness, mental retardation; muscular dystrophy, multiple sclerosis; drug addiction and alcoholism; visual, hearing, and communication disorders. This comprehensive Act affects state and local government, education, transportation, housing, and employment.

The Americans with Disabilities Act (ADA) of 1990 (Public Law 101-336) is considered a "civil rights bill" for individuals with disabilities. The ADA grants civil rights protection to individuals with disabilities in all public services, public accommodations, transportation, and telecommunications. Under ADA, employers with 15 or more employees may not refuse to hire or promote a person with a disability when that person is the most qualified person to perform the job. Additionally, an employer must make reasonable accommodation for a person with a disability if that

Figure 3.2
A "Barrier-Free Environment" is guaranteed by the Rehabilitation Act of 1973.

accommodation will allow the person to perform the essential functions of the job. Public accommodations affected by the ADA include auditoriums, restaurants, hotels, stores, banks, doctors' offices, museums, libraries, parks, zoos, schools, and recreation facilities. According to the ADA, it is discriminatory to fail to remove structural, architectural, and communication barriers in facilities where such removal is readily achievable. Telecommunications reform under ADA requires phone services to provide "Telecommunications Relay Services" for those with hearing or speech impediments.

Education. Local, state, and federal laws insure educational opportunities, provide for appropriate education, and seek to secure access to public buildings and streets for every child (see Figure 3.3). One famous decision was *Brown vs. the Board of Education of Topeka, Kansas* (1954), in which the court determined that racially segregated schools were separate but not equal in terms of educational opportunities for all students.

The Individuals with Disabilities Education Act (IDEA), formerly the Education of All Handicapped Children Act, became law in 1990. Its purpose is to guarantee the availability of special education programs to children and youth with disabilities and to assure that educational decisions relating to such students are fair and appropriate. Additionally, it assists state and local governments in providing special education through the use of federal funds. It requires schools to provide appropriate elementary and secondary education to children and youth ages 3-22. This public law has been amended several times and now includes the provision of services for adolescents who are leaving school programs and beginning adult life, and requires the educational program of a student with a disability to include a plan for transition to post-secondary life. This plan must include appropriate assistive technology which benefits the student, such as microcomputers, alternative speech devices, and keyboards.

Figure 3.3
Educational opportunities are guaranteed to every child.

Equal Protection. You may have been the well-cared for child of a prosperous parent whose economic and social status offered you many benefits. As a young adult, however, you may be an unemployed student with no health insurance and a social profile which makes you vulnerable to the scrutiny of law-enforcement officers. Later, you may become part of any number of minority groups, either because of your gender, sexual orientation, physical or mental disability, health, or religion. If you live many years, you will certainly be a senior citizen. Even the most fortunate of us will at some time in our lives come to realize the importance of equal protection.

Equal Protection of Laws is your Constitutional right to receive the same protection under state law as any other person. The 14th Amendment to the Constitution says that "No State shall ... deny to any person within its jurisdiction the equal protection of the laws." This clause was meant to prevent state governments from favoring particular groups of people at the expense of other groups. All people are protected, including the poor, children, prisoners, non citizens, and minorities. State constitutions also contain equal protection clauses.

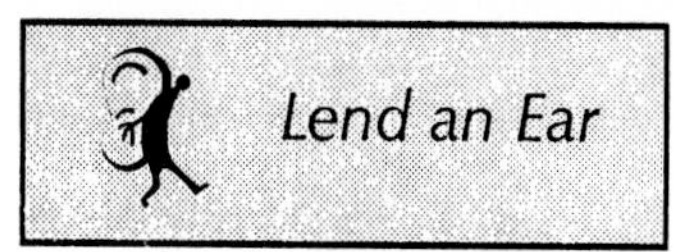

"The people's good is the highest law." —Cicero, Roman statesman

Families. The public policy decisions which have the greatest impact on children and families are made at the state and local levels. Child advocacy organizations speak out for improvement in the condition of the children in their community and state through legislative, executive, and judicial decisions on programs to meet children's needs. They focus on basic income and family support, child welfare, juvenile justice, nutrition, education, and child care programs. The term "family" does not have a precise legal meaning, so most laws include a definition of the term when they use it. In some legislation, the term means only people who are related by blood or marriage and live together. In other laws, family includes relatives who may or may not live

How many people are necessary to make up a legal family?

in the same household. In still other laws it can mean two people or a group of people who stay committed to one another in a domestic setting for an unlimited period of time, sharing a home and responsibility for financial support and household duties. No specific number of persons is necessary to make up a legal family (Leonard, 1990).

Gender. Ironically, such leaders as English Prime Minister Margaret Thatcher, Indian Prime Minister Indira Gandhi, and Pakistani Prime Minister Benazir Bhutto are often seen as anomalies rather than cited as models for gender equality. Though women have served as senior officers on space shuttles, supreme court justices, chief executive officers of major corporations, and at the highest ranks in the military services, their ability to lead is still seen as a subject of debate. Based upon their exemplary performance in traditional male roles, from combat commander to truck driver, it must be concluded that the debate stems from gender discrimination alone.

Though women are guaranteed equality, many women in the United States are still fighting to claim their full equal rights under the law. Women's advocates seek the passage of an Equal Rights Amendment and constantly monitor legislation affecting women both in the workplace and at home. Title VII of the 1964 Civil Rights Act prohibits sexual discrimination in employment. The Equal Pay Act of 1963 prohibits discrimination in the form of different compensation for jobs with equal skill and responsibility. The Pregnancy Discrimination Act of 1978 prohibits discrimination against employees on the basis of pregnancy and childbirth with respect to employment and benefits.

Race. In one generation we have seen extraordinary progress toward full integration and equality for all races and nationalities. For example, Charlayne Hunter (Gault) became familiar to television viewers across

America when, as a child, she had to be protected by the National Guard from protesters as she integrated a southern school. Today, Ms. Gault is widely known as a national television news anchor.

People of all different ethnic, racial, and national origins now have legal protection from discrimination (Coughlin, 1993). The 14th Amendment to the Constitution states that it is illegal to classify a citizen by his or her national origin, since a foreigner is entitled to equal protection under the law once that individual has been naturalized as a citizen. Additionally, Title VII of the 1964 Civil Rights Act prohibits discrimination based upon an employee's race, color, or national origin.

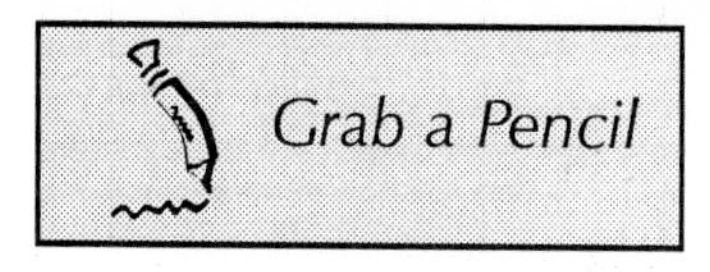

Can you think of some examples, either personal or public, which illustrate changes in our national attitudes regarding racial discrimination?

Religion. The U.S. Constitution guarantees that all Americans are free to follow the religion of their choice or none at all. Article I of the Bill of Rights states that "Congress shall make no law respecting an establishment of religion, or prohibiting the free exercise thereof." The government can neither establish a state religion nor force anyone to attend or support any religious institution. Individuals have the right to worship as they please. Title VII of the 1964 Civil Rights Act prohibits discrimination on the basis of an employee's religion.

Although the letter of the law may be followed, as individuals we need to be aware of the spirit of the law. Looking at the wars and dissension around the world in such places as the Middle East, Northern Ireland, and the states of the former Soviet Union, we may congratulate ourselves on our rights to religious freedom. However, though protected by laws, many of us continue to suffer from a subtle bias against certain religions. It is important to be sensitive to the fact that children, for instance, may feel this discrimination when certain holidays are celebrated in school. For example, many public schools in the United States celebrate Christmas in some way, if only by taking a school vacation, while Jewish, Moslem, Buddhist, Hindu, and other minority religions may be hardly acknowledged, if at all. Such

What might be some reasons for securing protection under the law for people who are homosexual?

oversight has a significant effect on the way our children perceive the importance of these religions.

Sexual Orientation. As of the publication of this textbook, no well-defined body of law guarantees the rights of people who are homosexual. No federal law protects gay men and lesbians from discrimination in employment or housing. Advocacy groups continually focus on combating state-by-state anti-gay ballot initiatives, challenging the military ban, and establishing legal precedents to secure constitutional protections for lesbians and gay men.

As attorney and civil rights expert Donald Cohen explains, most anti-homosexual laws treat gay men and lesbians as a class of criminals who are not entitled to assert their civil rights. In many states, private consensual sexual acts between same sex adults remain criminal, and individuals have been imprisoned simply for expressing their sexuality. Such anti-homosexual laws mistakenly equate being gay, or living an alternative lifestyle, with the criminal act of sodomy. As long as being gay is considered criminal, Cohen suggests, there can be no secure and lasting legal rights for all people (Cohen, 1994).

Cohen considers criminal laws which proscribe consensual sexual activity to be the central obstacle to homosexual rights. In Bowers v. Hardwick (1986), the United States Supreme Court held that states may criminalize sexual activity between consenting adults. "The right to privacy, therefore, did not protect people from being arrested for private sexual conduct even in their own homes," Cohen explains. Yet the battle for homosexual rights on state constitutional grounds continues to be an evolving process, he says, and the tide is turning. For example, in Texas v. Morales (1992), an appellate court held that the Texas State Constitution protected the right to choose one's sexual orientation. And in Kentucky v. Wasson (1992), the Kentucky Supreme Court decided that the Kentucky Constitution protects the freedom to choose one's own sexual partner and struck down the state's anti-sodomy law (Cohen, 1994).

Because our laws still don't protect equal rights on the basis of sexual orientation, people who are gay, lesbian, or bisexual continue to face discriminatory practices. It should also be recognized that people who are heterosexual also have no legal protection on the basis of their sexual orientation.

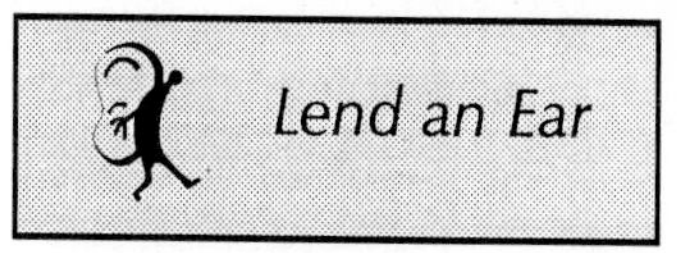

"I am the inferior of any man whose rights I trample under foot." —Robert Green Ingersoll, political leader

Size. As of the publication of this textbook, no well-defined body of law guarantees the rights of people who are overweight or excessively tall or short. Advocacy groups seek to fight discrimination against people on the basis of weight or stature. They work to educate lawmakers and serve as national legal clearinghouses for attorneys challenging size discrimination. Our media culture clearly promotes certain physical attributes such as tallness and thinness. Perhaps because height is perceived as a positive attribute, there seems to be little awareness of the challenges of those who are exceptionally tall. The inaccessibility of public facilities is also problematic for those who may not find ample legroom, headroom, or seating space in public transportation, theatres, schools, and other such places.

Socioeconomic Levels. Socioeconomic disadvantage is a complex, multi-faceted problem for many Americans of both genders and all races and national origins. The Equal Protection Clause of the 14th Amendment guarantees equal protection under the law for people living in poverty (see Figure 3.4). Federal Social Security programs provide aid to families with needy dependent children, supplemental security income to persons with disabilities, and Medicaid to ensure health care for low-income families. The Commodities Distribution Act distributes surplus staple foods to needy families, and the Food Stamp Act of 1904 distributes coupons that may be exchanged for food. Advocacy groups monitor state and

Figure 3.4
People living in poverty are guaranteed equal protection under the law.

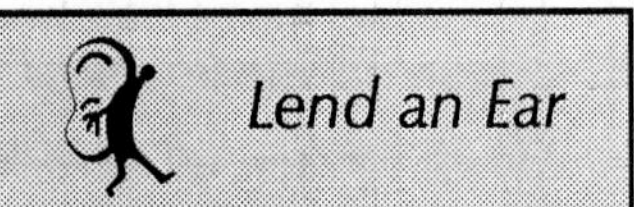

"Freedom really is indivisible. Truth does not need the 'protection' of any bureaucrat, librarian, or Webmaster. Indeed, truth becomes most clear and powerful when allowed to shine side-by-side with all of its competitors. However politically correct or seemingly well-intentioned, 'protected truth' always putrefies into oppressive dogma." —Editorial in *The Guide* magazine of Boston

local legislation, focusing on welfare reform, job creation, public housing, and job training.

The Job Training and Partnership Act (JTPA) of 1982 (Public Law 97-300) authorizes federal funds administered by the Department of Labor to support local and regional employment assistance efforts. The JTPA offers opportunities to address such diverse elements as basic literacy, skills development, and education in life skills such as job interviewing, budgeting, and parenting. For example, it offers low-income or unemployed single parents the chance they may need to find financial aid, child care assistance, and parenting and legal advice so that theycan address their education needs and learn a skill or trade.

Bringing About Change

As we have become educated as a society about the causes and effects of exceptionalities, we have made some significant, though still inadequate, progress in the development of legal protection of human rights, and through that legislation have begun to undergo real social change. Countless individuals have sacrificed time, money, energy, and sometimes their lives to bring about these changes. For example, civil rights leaders Martin Luther King, Jr. and Madger Evers gave their lives for their causes. Rosa Parks risked her life for integration of public transportation. Col. Margarethe Cammermeyer revealed her sexual orientation at the risk of her military career. Vietnam war veteran Ron Kovic dedicated his life to securing the rights of veterans with disabilities. Madeline Murray O'Hair, an atheist, fought for her freedom *not* to worship. Many challenges remain in the attempt to secure full participation in society for each and every person regardless of race, gender, religion, sexual orientation, ethnicity, age, or disability. It will take the continued efforts of dedicated individuals to bring about more change (see Figure 3.5).

Figure 3.5
Only the efforts of dedicated individuals can manifest improved legal protection of human rights.

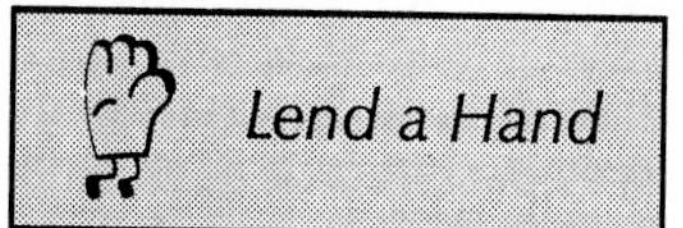

Assist someone who needs help in understanding his or her legal rights.

Conclusion

As time passes, new issues arise which demand attention and resolution. In the 1980's and 90's, a major new source of concern was the question of how to meet the needs of people with AIDS and other contagious diseases while protecting their privacy and insuring their rights. Other on-going issues center around the problem of drug abuse as it has an impact on the society at large. New technology brings new possibilities for treatment of personal and social ills, but the paramount issue is still the preservation of human rights.

References

Cohen, D. G. (1994). The rights of gay men and lesbians. *Legal Interest,* 5-10.

Coughlin, G. G. (1993). *Your handbook of everyday law.* New York: Harper Perennial.

Leonard, R. (1990). *Family law dictionary.* Berkeley, CA: Nolo Press.

Suggested Readings

Curry, H. (1994). *A legal guide for lesbian and gay couples.* Berkeley, CA: Nolo Press.

This is a self-help guide for gay couples which covers various legal statutes affecting homosexuals.

Gifis, S. H. (1991). *Law dictionary.* New York: Barrons.

This is an indispensable handbook covering United States law and defining many important legal terms.

Leonard, R. (1990). *Family law dictionary.* Berkeley, CA: Nolo Press.
Provided in this book are some easy to understand definitions of legal terms and legislation important to families in the United States.

U.S. Deptartment of Justice. (1990). *The Americans with disabilities act.* Washington DC: U.S. Department of Justice.
This pamphlet presents a brief overview of the Americans with Disabilities Act.

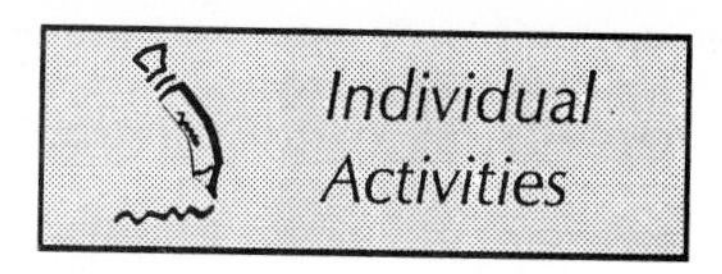

1. Do you think that most citizens know about the laws concerning human diversity and their implications? Conduct an informal survey of your friends and family.

Observations:

2. As you go through your daily activities, you often notice the accommodations made for people with disabilities–ramps, lower water fountains, Braille on elevator buttons, etc. Imagine that you are a politician who would like to design a bill to help members of another category of diversity. Describe legislation that you would sponsor to better the quality of life for diverse people.

1. Under Equal Protection of Laws, all people are protected, including prisoners and non citizens. Discuss in your group whether or not you believe that prisoners and non citizens are entitled to equal protection.

Reactions:

2. A bill is being debated concerning the rights of excessively tall individuals. This bill, if passed, will grant tall people free college education and low cost housing. Your class is the deciding body for this law. Choose six members of the class to make the final decision based upon the arguments presented by the rest of the class. Arrange to have half the class argue the pros of this bill and the other half the cons.

Reactions:

Reflection Paper 3.1

The Americans with Disabilities Act is a civil rights bill for people with disabilities. It states that the workplace must be made to accommodate people with disabilities. Some employers don't support the ADA. For what reasons may they have a negative attitude toward the ADA? What counter-arguments would you offer to them?

Reflection Paper 3.2

Some people believe that laws are needed to improve the quality of life for persons who are gay or lesbian. What is your position? Why?

Section II:

Categories of Human Diversity

Chapter 4 Racial and Ethnic Diversity

Chapter Sections

- Introduction
- The Culture of Race and Ethnicity
- Terminology
- Are You a Racist?
- Racial Diversity
- The Problem with Racial Classification
- The Uses of Racial Classification
- A Changing Definition of "Minority"
- Support Services
- Interactions
- Conclusion
- Activities

"Men are not superior by reason of the accidents of race or color. They are superior who have the best heart—the best brain."
—Robert Green Ingersoll, political leader

Objectives

- Explain how race and ethnicity are distinguished.
- Specify how the U. S. Census Bureau defines race.
- Distinguish between scapegoating and discrimination.
- Identify two factors responsible for blurring racial distinctions.

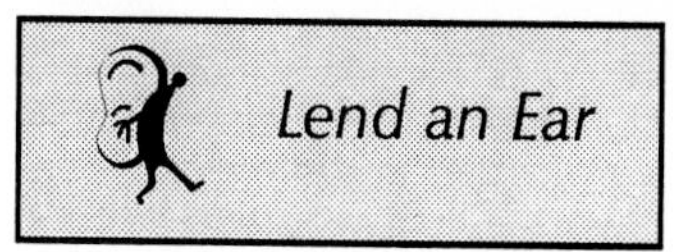

"It's time for us to turn to each other, not on each other." —Rev. Jesse Jackson, civil rights leader

Introduction

More than any other category of human diversity, the concept of "race" is controversial and explosive. Among both politicians and humanitarians there is a growing perception that it is desirable to minimize racial differences and to emphasize human similarities. Some anthropologists have even gone so far as to state that there is no such thing as race at all, maintaining that the study of human differences merely serves to perpetuate prejudice (May, 1971). Yet interracial tensions and hostilities persist, and many people feel that minimizing differences is done at the risk of sacrificing important racial and ethnic distinctions celebrated by various groups (see Figure 4.1).

It seems pointless to deny the existence of different types of people, especially since unity and harmony among peoples do not require that everyone be identical. In fact, the definition of harmony is many *different* voices in accord. The purpose of this chapter, then, is to define the concepts of race and ethnicity in the hope that an understanding of them will encourage a genuine brotherhood and sisterhood among humans. You will have the task of examining your own feelings and prejudices as you cultivate fair and objective attitudes.

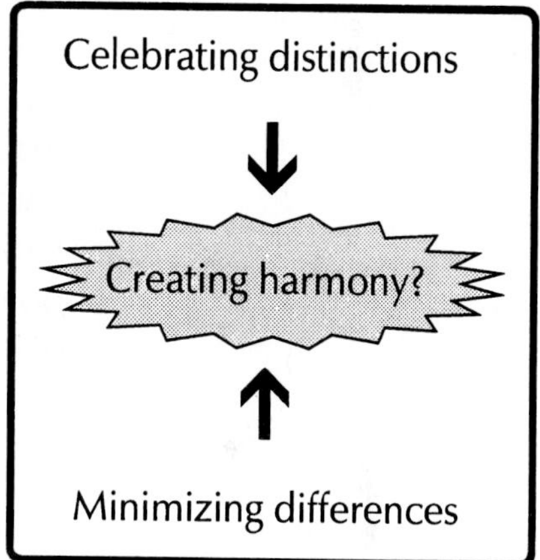

Figure 4.1
There is a delicate balance between avoiding stereotypes and recognizing differences.

The Culture of Race and Ethnicity

It is not possible to make any definitive statements about the culture of race beyond the fact that culture and race are generally considered to be separate issues. Much of what we associate as racial culture is actually geographic and social in nature. A person's country of origin and his or her race are distinct. People of African origin, for example, may be of any race. Likewise, how we define "African culture" depends upon where in the world we look. The people of the Congo share much cultural heritage with the French, the people of Somalia are culturally defined by their Muslim heritage, and the African American population has created its own unique culture influenced by America's diverse immigrant population. With exponentially increased immigration and travel worldwide, distinctions of race and culture are blurring (see Figure 4.2).

Figure 4.2
Global communications, travel, and immigration are helping to blend racial and cultural differences.

Ethnicity, on the other hand, can definitely be linked to culture. In fact, ethnicity arises from the cultural commodities that are shared by a number of individuals (Auerbach, 1994). Such commodities are magnets that draw individuals together, providing them with a group identity and consciousness. A shared nationality, language, religion, or history may be the focal point of an ethnic group. Like races, ethnicities often overlap. A person with parents of Cuban and German backgrounds could choose to be called either Cuban American, German American, or simply American, and could celebrate any one cultural heritage or all three.

The challenge for society is to focus on human attributes, while keeping aware of and honoring ethnic differences. Sometimes people are actually prevented from building bridges over ethnic and racial boundaries. A typical boundary occurs when one ethnic group perceives another as not being equal. In other words, one group values certain qualities in the other more highly than in its own. Psychologists suggest that members of ethnic groups "may pay particular attention to differences between themselves and others in order to bolster a positive sense of their social identity" (Bower 1996, p. 409). Therefore, individuals in two different groups may be prevented from connecting with each other on "neutral ground" (Garcia 1994, p. 184).

Consider This

Some anthropologists suggest that the study of human differences merely serves to perpetuate prejudice. Do you agree or disagree? Why?

Terminology

Before we delve into the topic of racial and ethnic diversity, an understanding of the following terms is crucial.

Americanization. This term refers to a practice of acculturation that seeks to merge small ethnic and linguistically diverse communities into a single dominant national culture (Garcia, 1994). Americanization is a synonym for the "melting pot" phenomenon.

Bias. This is a personal preference that prevents one from making a fair judgment.

Bigotry. This is a stubborn intolerance of any race, nationality, or creed that differs from one's own.

Discrimination. This is differential treatment based on unfair categorization. It is denial of justice prompted by prejudice. When we act on our prejudices, we engage in discrimination. Discrimination often involves keeping people out of activities or places because of the group to which they belong.

Ethnicity. This refers to how members of a group perceive themselves and how they are in turn perceived by others. *Ethnic* describes a group of people within a larger society that is socially distinguished or set apart by others and/or by itself, primarily on the basis of racial and/or cultural characteristics. Such characteristics may include religion, language, and tradition. The beliefs of individuals about their own ethnic group tend to be similar to, and more positive than, the beliefs of those outside the group.

Nationality. One's nationality is not necessarily tied to one's culture, ethnicity, or race. Nationality simply refers to the country where a person was born.

Minority. Although this term should simply be used to describe a subset within a population, it often connotes inferior or lesser status in comparison to the majority. Often the minority is a sociological term referring to a social group that occupies a subordinate position in a society.

Prejudice. This is a set of rigid and unfavorable attitudes toward a particular group which is formed in disregard of facts. Prejudice is usually an unsupported judgment accompanied by disapproval. This is a learned concept, not innate. Individuals generally do not realize how prejudiced they actually are.

Figure 4.3
Race is about superficial characteristics only, such as the shape of the lips, nose, eyes, and forehead.

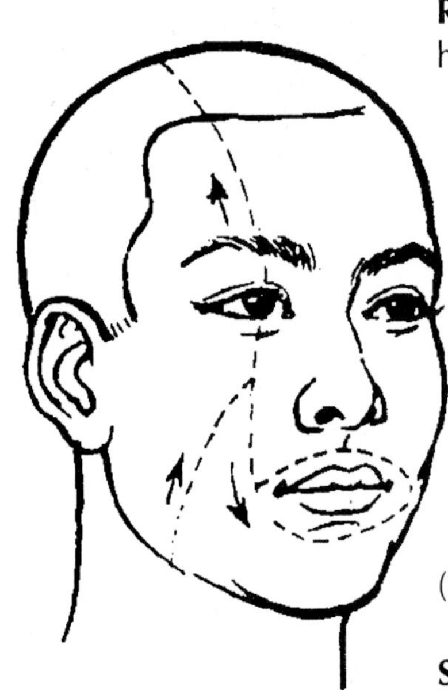

Race. This is an anthropological concept used to divide humankind into categories based on physical characteristics of size and shape of the head, eyes, ears, lips, and nose, and the color of skin and eyes (Bennett, 1993). Race has never scientifically been equated with any mental characteristics such as intelligence, personality, or character (see Figure 4.3).

Racism. This is the belief that one's own race is superior to another. This belief is based on the erroneous assumption that physical attributes of a racial group determine their social behavior as well as their psychological and intellectual characteristics (Bennett, 1986).

Scapegoating. This refers to the deliberate policy of blaming an individual or group when the fault actually lies elsewhere. It means blaming another group or individual for things they did not really do. Those that we scapegoat become objects of our aggression in work and deed. Prejudicial attitudes and discriminatory acts often lead to scapegoating. Members of disliked groups are denied employment, housing, political rights, or social privileges. Scapegoating can lead to verbal and physical violence, including death.

Have you ever been blamed for something you did not do? Have you ever been the victim of scapegoating?

Stereotyping. This is a preconceived or oversimplified generalization involving beliefs about a particular group. Negative stereotypes are frequently the foundation of prejudice. The danger of stereotyping is that it ignores people as individuals and instead categorizes them as members of a group who all think and behave the same way. We may pick up these stereotypes from what we hear other people say, what we read, and what people around us believe. As a group becomes more distinctive in character, a consensus develops regarding stereotypes associated with that group (see Figure 4.4).

Tolerance. This is the opposite of bigotry. Tolerance refers to a fair and objective attitude toward races, nationalities, and practices different from one's own. However, tolerance implies disapproval. Tolerance does not celebrate diversity but merely puts up with it.

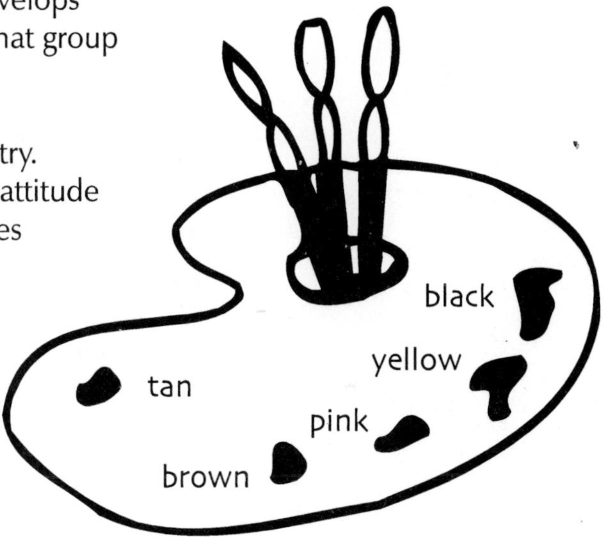

Figure 4.4
Racial stereotypes, typically involving skin color, ignore people as individuals, instead lumping them into general groups.

Are You a Racist?

The following self-test will help you to assess your own thoughts and feelings about the issues of racial and ethnic diversity. We have accumulated concepts and attitudes throughout our lives, some of which we may never have consciously examined. Answer yes or no to each of the following

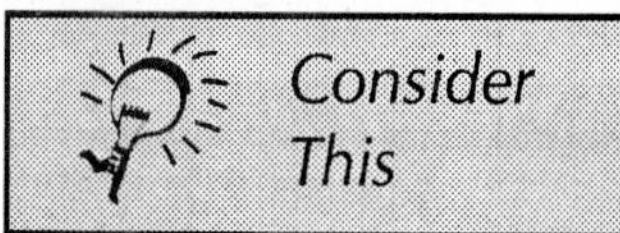

We pick up racial stereotypes from what we hear other people say, what we read, and what people around us believe. What are some of your own stereotypes, and how did you learn them?

questions (adapted from Patterson & Kim, 1991):

1. Do you believe that your country would have fewer racial problems if different people would keep to their own kind?
2. Would it bother you to learn after an accident that you received a blood transfusion from someone of a different race?
3. Have you ever invited a person of another race into your home for a social occasion?
4. Would your friends be understanding if you married a person of another race?
5. Do you believe that some races are naturally superior to others?
6. Would you be reluctant to adopt a child of a different race?
7. Do you think that people of some races are more intelligent than others?
8. Do you think people of some races are more hard-working than others?
9. Do you think legal immigrants are stealing American jobs?
10. Would you send your child to a private school that did not accept racially diverse students?
11. Do you feel uncomfortable around people of different races?
12. Does it make you uneasy to see a person holding hands with someone of a different race?
13. Do you have any racially diverse friends?
14. Do you believe that so-called minorities are poor because they fail to take advantage of the opportunities open to them?

Give yourself one point if you answered "yes" to questions 1, 2, 5, 6, 7, 8, 9, 10, 11, 12, 14. Give yourself one point each if you answered "no" to questions 3, 4, 13.

How did you score?

9 - 14 = *stubborn racist*
(you categorize and judge people based solely upon race)
7 - 8 = *bigoted*
(you are generally intolerant of any race that is not your own)
5 - 6 = *prejudiced*
(you have preconceived ideas about people based on race)
3 - 4 = *biased*
(your personal preferences color some of your judgments)
2 = *tolerant*
(you grant others their rights but still see differences based upon race)
0 - 1 = *color-blind*
(you do not categorize or judge people based upon race)

Racial Diversity

Racial differentiation is a product of evolution, natural selection, and genetic mixing, and human races vary greatly across the globe. As Edward Babun pointed out in *The Varieties of Man,* people of genetically diverse populations look different in superficial ways, such as skin color, facial features, body build and size, and hair type. They also differ in less apparent ways, such as resistance to various diseases, tolerance for extreme temperatures, frequencies of blood type, and other aspects of biochemistry and physiology (see Figure 4.5). For all the external and internal differences, however, human races ultimately make up one species, *Homo sapiens* (Babun, 1969).

In the gene pool of any given race is an astronomical number of possible chromosome

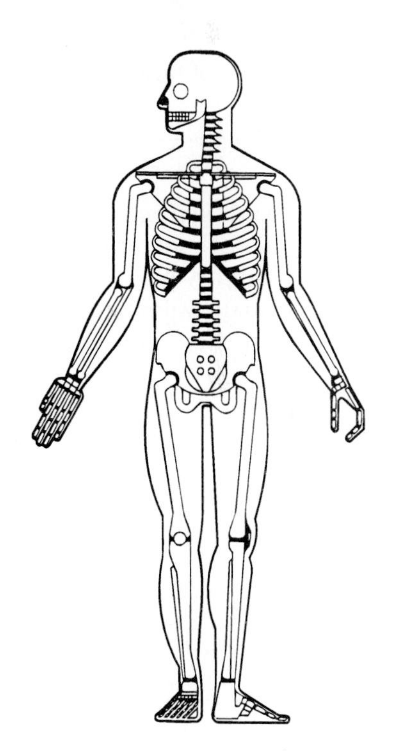

Figure 4.5
Genetically diverse populations may differ in terms of biochemistry and physiology.

Ask a friend to solve this brain teaser:

"Which pair has more genetic differences?"

Pair A
a native of China and a native of Australia

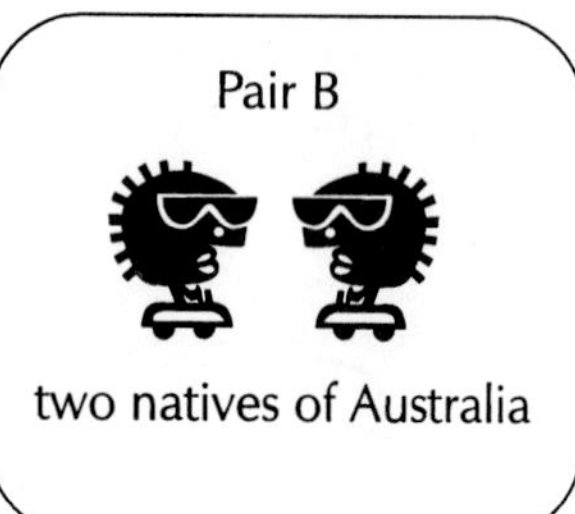

Answer: Pair B is more genetically diverse, since individual variation within local groups accounts for 85% of genetic difference.

combinations. Differences among races account for a mere six percent of human genetic variation. Differences among tribes or nations within a race account for only eight percent of variation. Individual variation within local groups, on the other hand, accounts for eighty-five percent of genetic difference (Myers, 1995). Therefore, the average genetic difference between two Siberians or two Cubans is much greater than the difference between the two groups. Not all Japanese people are short with black hair. Not all Norwegians are tall with blond hair. In every race, the range of difference is vast.

The Problem with Racial Classification

The actual number of races has not been clearly defined. Some scientists classify humankind into five major stocks or subspecies (May, 1971):

- **Mongoloid:** yellow, tan and copper-colored Asian people with narrow or almond-shaped eyes, straight black hair, and wide cheekbones.
- **Negroid:** tan, black and brown-colored African people with tightly curled brown or black hair, high foreheads, and large lips and noses.
- **Capoid:** yellow-tan colored South African people with narrow eyes.
- **Caucasoid:** pinkish-white and olive-colored European people with blond to brunette and wavy to straight hair, blue or green eyes, and straight or pug noses.
- **Australoid:** dark brown to black-colored Australian, Indian, and Southeast Asian people with curly or wavy hair and heavy browbridges.

Other scientists consider the traditional five groups to be outdated and have pinpointed as many as two thousand racial populations. The reason for such disagreement is that human races are constantly evolving and changing. Some of today's living races did not exist five hundred years ago, and others are older than written history (Babun, 1969). There hasn't been a pure race for more than 100,000 years (Duvall, 1994), and not all racial populations are completely distinct.

It is common today to find people who have backgrounds in two or more races, making it even more difficult to classify them. In fact, the very concept of race, according to the Census Bureau, now reflects self-identification and does not denote any clear-cut scientific definition of biological stock. How one defines his or her race ultimately depends upon which race he or she most closely identifies with (Famighetti, 1993).

The problem with racial classification goes deeper than merely what categories or how many categories we choose to use. Geneticists have had trouble even identifying the gene or sets of genes responsible for race, and many now claim that race does not exist.

The idea of race is actually relatively new in human history. The term was coined by the French naturalist George Louis Leclerc Buffon in 1749. Perhaps the concept could disappear just as quickly as it was invented. "If members of society refused to believe that skin color and certain other physical traits were important, the concept of race would not exist" (Auerbach, 1994, p. 1377).

The Uses of Racial Classification

In spite of the fact that racial classification is virtually meaningless, our society categorizes people by race all the time. We use racial labels on birth certificates, marriage licenses, college applications, and other official forms. Race is used to determine eligibility for education

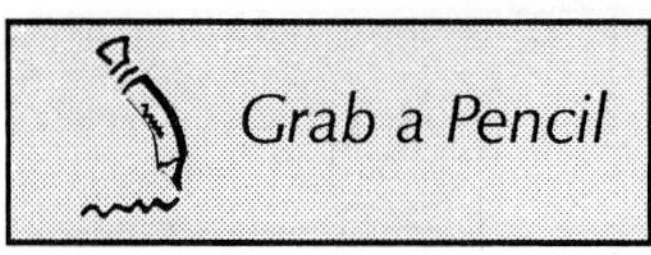

Imagine a world in which the concept of race did not exist. Brainstorm five ways in which that world would be a better place.

My son is adopted. We brought him to the U.S. from Korea. When he was old enough, I talked to him about the fact that he was born in another country but that now he was American.

Shortly after he started kindergarten, he began to resist going to school. When I questioned him, he said the children called him names because he is Chinese.

"But that's silly," I told him. "You're not Chinese, you're Korean." What he said made me realize that I had no way to help him understand racism, because I couldn't understand it myself. "But Mom," he said, "they don't know any names for Koreans, so that's why they call me a Chink." He had already accepted the idea that racism was a given, and he was only 5.
—Dinah L., San Diego, CA

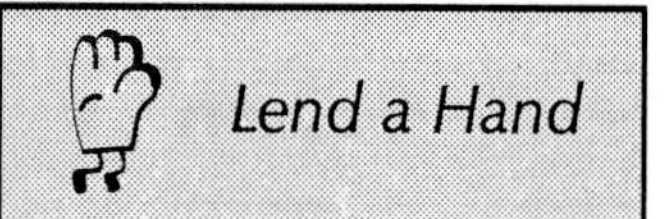

When was the last time you heard a racist joke or remark? What was the context? What do you think led the person to make such a remark in the first place? What was your reaction, and why did you react the way you did? Will you react differently in the future? If so, how?

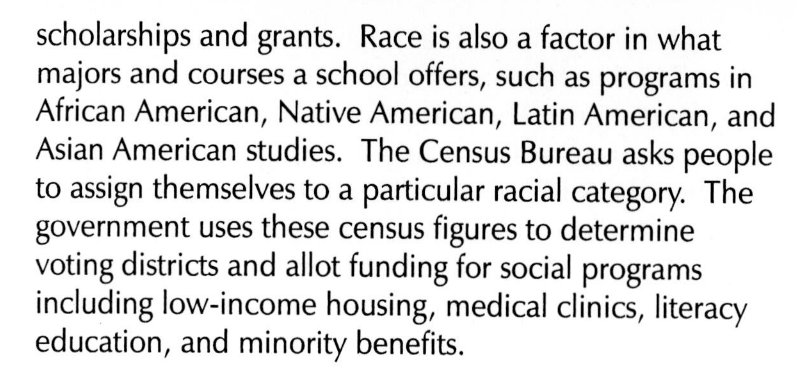
scholarships and grants. Race is also a factor in what majors and courses a school offers, such as programs in African American, Native American, Latin American, and Asian American studies. The Census Bureau asks people to assign themselves to a particular racial category. The government uses these census figures to determine voting districts and allot funding for social programs including low-income housing, medical clinics, literacy education, and minority benefits.

A Changing Definition of "Minority"

Just as our conception of what "race" means is changing, our definition of "minority" must change as well. That's because the demographics in the United States are shifting rapidly. In 1995, nearly one out of every four Americans was a member of a so-called minority racial or ethnic group. Estimates of future population predict that after the year 2000 one out of every three Americans will be a minority (see Figure 4.6). By 2050, the percentage of the population that is White (not Hispanic) will decrease from 75.7 percent (in 1990) to 52.7 percent (Landes, 1994). With each passing day, the word "minority" loses both its original meaning (a smaller part) and its connotation (something lesser than the majority). We are in the process of refining and redefining what we mean when we use racial categories. The very fact that we are examining our old ideas offers hope of a more enlightened attitude.

Figure 4.6
As society becomes more diverse with each passing year, the definition of "minority" is changing.

Support Services

Ethnic and racial populations have support groups, churches, and social organizations where people can interact and share their common heritage. Members meet and communicate through newsletters and social opportunities. Affirmative Action policies, which are often the target of debate, assist people who are discriminated against because of their race to obtain equal treatment in housing, employment, and banking.

Interactions

Take a stand against racist behavior. Don't be a guilty by-stander. Remaining silent only makes prejudiced people think you agree with them.

Remove "race" from your vocabulary. There is no scientific definition for race. It is a vague, imprecise category. The concept of race only serves to promote divisiveness.

Practice looking at people and not at labels. It often takes a conscious effort to unlearn prejudices and biases we have grown up with (see Figure 4.7). Since members of any given group are just as diverse as members of any two groups, *everybody* is an "other."

Treat everyone you meet as an equal. Superficial differences have nothing to do with intelligence or ability. Beneath every skin tone and hair color is a human being just like you, experiencing many of the same joys, worries, doubts, and challenges inherent in being alive. Take differences at face value.

Celebrate a particular ethnic culture. Remember that ethnic groups are made up of people who *choose* to associate with each other. Your ethnicity is whatever you make it, and the celebration of your ethnic group tends to have a positive impact on everyone's lives.

Accept, respect, and celebrate diversity. Different individuals possess unique talents and personalities, and those should be respected, celebrated, and indeed sought after because they enrich all our lives.

Figure 4.7
What do you see in the picture above? Is it a Native American? Practice looking at the *person* beneath the label.

Untrue Stereotypes of People from Other Countries

- They are in the United States illegally
- They spend all their money in their home country
- They do not understand English
- They have a large family
- They have a lot of money or come from a wealthy family
- They are stealing American jobs
- They are either work-a-holics or lazy

Issues Racially Diverse People May Face

- Prejudice and bigotry
- Being labeled as a foreigner
- Being treated differently
- Homesickness, separation from family
- Ignorance about country of origin or heritage
- Being perceived as an illegal immigrant
- Being perceived as "Un-American"
- Cultural differences
- Overcoming stereotypes
- Interracial relationships
- Finding peer support
- Hate crimes

Conclusion

As members of the genetically diverse human race, we need to appreciate our similarities as well as our differences. Skin color is the most common factor people use to separate themselves into smaller groups and "races." However, skin color is the perfect example of how race is relative. Depending upon what society or circumstance you are in, the shade of your skin may be variously interpreted. For example, the Creoles of Louisiana, a French-speaking people of mixed racial heritage, are variously seen as black, white, or Creole, depending upon the context (Auerbach, 1994). Perhaps we cannot ignore our superficial, genetic differences, but we should strive to take such differences at face value, so to speak. Ultimately, we are all members of the *human* race.

References

Auerbach, S. (Ed.). (1994). *The encyclopedia of multiculturalism.* (Vol. 5). New York: Marshall Cavendish.

Babun, Edward. (1969). *The varieties of man: An introduction to human races.* London: Collier-Macmillan.

Bennett, C. E. (1993). The black population in the United States. *Current Population Reports, 20-471.*

Bennett, C. I. (1986). *Comprehensive multicultural education: Theory and practice.* Boston: Allyn and Bacon.

Bower, B. (1996, June 29). Fighting stereotype stigma: Studies chart accuracy, usefulness of inferences about social groups. *Science News, 149,* 408-409.

Duvall, Lynn. (1994). *Respecting our differences: A guide to getting along in a changing world.* Minneapolis: Free Spirit.

Famighetti, R. (Ed.). (1993). *The world almanac and book of facts 1994.* Mahwah, NJ: Funk and Wagnalls.

Garcia, E. (1994). *Understanding and meeting the challenge of student cultural diversity.* Boston: Houghton Mifflin.

Landes, A. (Ed.). (1994). *Minorities: A changing role in America.* Wylie, TX: Information Plus.

May, J. (1971). *Why people are different colors.* New York: Holiday House.

Myers, D. G. (1995). *Psychology.* New York: Hope.

Patterson, J. & Kim, P. (1991). *The day America told the truth.* New York: Plume.

Suggested Readings

Baker, J. R. (1974). *Race.* New York: Oxford University Press.
This book defines the concepts of race and ethnicity and reviews evidence of the intelligence of four different human races.

Landes, A. (Ed.). (1994). *Minorities: A changing role in America.* Wylie, TX: Information Plus.
With up-to-date charts and diagrams, this book covers population trends that will force us to redefine what we mean by "minorities."

Thernstrom, Stephan. (Ed.). (1980). *Harvard Encyclopedia of American Ethnic Groups.* Cambridge: Harvard University Press.
This comprehensive book examines hundreds of categories of ethnic diversity, providing an informative, easy-to-read essay on each.

Notes

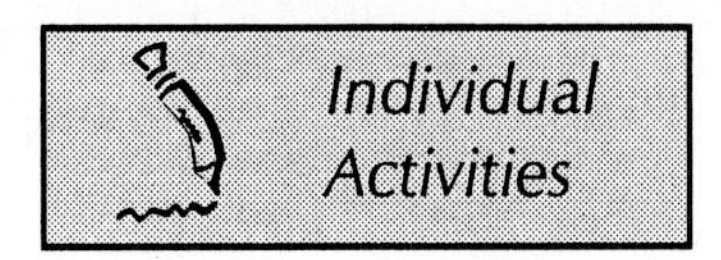

1. Thoroughly read today's newspaper. Then answer the following questions (adapted from Duvall, 1994):

a. How many stories involve racism or racial issues?

b. How many members of "minority" groups are mentioned?

c. How many stories are about members of the "majority"? (How can you tell?)

d. In what context are minorities mentioned? (Criminals? Victims? Experts? Prominent members of society?)

e. How often do members of minority groups appear in negative stories as opposed to positive ones?

Reactions:

2. How many stereotypes about your own race can you think of? Are any of them valid?

Reactions:

1. As a group, discuss what constitutes an "Average American." Where did your mental images of the Average American come from?

Reactions:

2. As a group, discuss how life will be different for white Americans when they are no longer the majority. How will life be different in the home, at school, at work, and in the community?

Reactions:

Reflection Paper 4.1

The definition of "minority" has been changing in the United States. In what ways will this affect future generations?

Reflection Paper 4.2

Do you think it is valuable for people to categorize themselves on official forms (such as job, loan, or scholarship applications)? Why or why not?

Notes

Chapter 5 Gender and Sexual Orientation

Chapter Sections

"Most times, the only gay or lesbian face people know of is who they see in the pride parade. To judge us on that would be like judging heterosexuals after watching Mardi Gras."
—Candace Gingrich, activist

Objectives

- Distinguish between sex and gender.
- Explain the best way to respond to sexist biases and stereotypes.
- Identify which labels are appropriate for describing sexual orientation.
- Describe the major issues addressed by the gender movements.

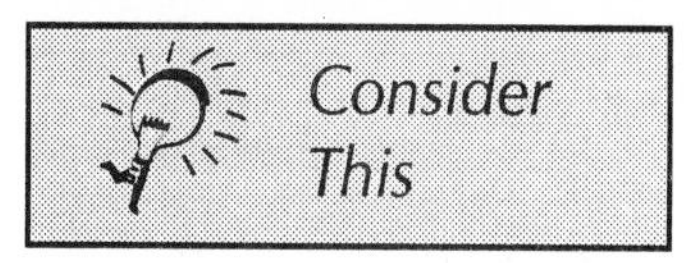

Our sexual orientation is invisible unless we choose to reveal it. In what ways do you reveal your sexual identity to society? Where do you draw the line on public displays of affection?

Introduction

Since the moment you were born and the doctor said, "It's a boy" or "It's a girl," your parents, peers, and society have helped to shape your sexual identity. Every human being, whether male of female, has gender and sexual orientation. As Schuman and Olufs (1995) have pointed out, it is important to remember that gender is not the same as sex. Sex is a biological classification, relating to specific organs of reproduction. Gender is a much broader topic which focuses on the roles occupied by males and females (see Figure 5.1). There is a question as to whether sex and gender are even related, and so we use the word *gender* to make the distinction (Schuman & Olufs, 1995). While members of minority groups often have characteristics which make them apparently different from the majority, such as skin color, language, age, or size, our sexual orientation is invisible unless we choose to reveal it.

Our sexual identity is so fundamental a part of who we are that it is a most difficult subject to discuss objectively. There are differing scientific views about gender differences and about the definition and roots of sexual orientation. However, everyone agrees on one point: the issues are complex and few people can address them without emotion. That is probably

Figure 5.1
A person's sex relates to the specific organs of reproduction. Gender is a separate issue which focuses on male and female roles.

because we are all sexual beings. The purpose of this chapter is to examine the diversity of human sexuality and to suggest ways we can combat discrimination.

The Culture of Sexuality

The sexual impulse is a powerful driving force of culture. It colors and enriches our lives. The idealization of male and female beauty has given us famous works of art, most notably the ancient Greek Venus de Milo statue and Michaelangelo's sculpture of David. The longing for human connection, both spiritual, emotional, and sexual, is the subject of countless love stories, plays, ballets, and operas. When we think of sexuality, it is important not to confuse it with pornography and perversion. Too often people think of all sexuality as something bad or dirty. To do that is to think of ourselves as unnatural and to ignore an integral part of our being.

Figure 5.2
Men and women are often not treated equally. The Gender Gap ignores the fact that women and men should be able to try anything.

The Gender Gap

The gender gap refers to inequality between the sexes (see Figure 5.2). For example, the life expectancy for American men is approximately eight years shorter than that for women. This life span disparity is in part due to the fact that men suffer from a variety of stress-related diseases which are connected to sexism. That is, men try to live up to societal expectations and ruin their health in the process (Hanmer, 1990). Another example of the gender gap is the fact that women in America earn about 70 cents for every dollar men are paid for similar full time employment. This wage disparity exists even though most women are self-supporting heads of households or necessary contributors to a family income. The gender gap arises out of a fundamental belief that men and women are not equal. The gender gap fuels sex discrimination.

Discrimination in Action

Prejudice and discrimination against people because of their gender or their sexual orientation is sometimes an even greater problem than that against other minorities because it potentially involves half the population of the earth. When we have discrimination against either men or women built into our laws and our system of education, for example, we may not even think of ourselves as being biased against one gender or the other. It is merely our way of life. While not everyone is a member of any other minority group, everyone has a particular gender and sexual orientation.

Whether knowingly or unknowingly, individuals and cultures practice sexual discrimination all the time. Masculine and feminine stereotypes define standards of beauty which dictate how men and women decorate themselves. Gender role stereotypes have led us to distinguish between "men's" jobs and "women's" jobs and have shaped our legislation for child custody (see Figure 5.3). Jokes, slams, slurs, graffiti, and whistling reveal our biased attitudes about gender and sexual orientation. Rapes and assaults carry sexist attitudes to a violent extreme. Whatever our sexual orientation, and whichever our gender, all of us suffer in some way every day because of our own lack of information and understanding about human sexuality. Fortunately, ignorance is a curable disease.

Untrue Gender Stereotypes

- Men are all sexists
- Men are emotionally strong
- Men never listen
- Men are not affectionate
- All men are homophobic
- Men are sexually aggressive
- Men are weak if they have feminine qualities
- Men and women are opposites

- Women cannot be sexist
- Women are emotionally weak
- Women always want to talk
- Women are all affectionate
- Women aren't homophobic
- Women don't care for sex
- Women are strong if they have masculine qualities
- Women and men are not different

(adapted from Papish, 1995)

Figure 5.3
Workplace discrimination has fueled protest movements.

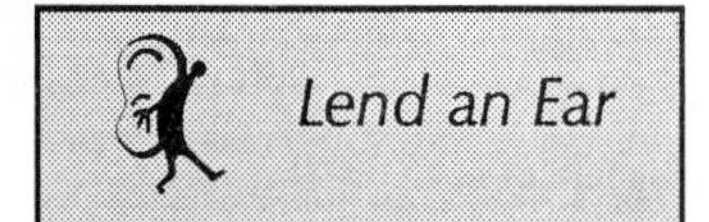

"We hold these truths to be self-evident, that all men and women are created equal."
—First Women's Rights Convention, July 19, 1848

Issues People May Face Related to Gender

- Lack of understanding of the other gender
- Sexual assault
- Stereotyped behaviors for men and women
- Stereotyped gender roles
- Developing relationships
- Puberty
- Cultural differences in defining masculinity and femininity
- Sexual harassment
- Equal opportunity
- Pressures to have a family
- Pressures to get married
- Pressures to have children
- Stereotyped careers
- Defining masculinity and femininity
- Body image
- Sexism in general
- Sexual behavior
- Finding role models
- Gender equity

(adapted from Papish, 1995)

The Gender Movements

Whether you are male or female, the way your gender is perceived and treated in the world is no doubt of great importance to you. Though citizens in the United States enjoy more freedom than in much of the world, the struggle for gender equity is ongoing. Sexual harassment, sexist institutions, and domestic violence continue to thwart our fight for rights. The women's and men's movements seek to empower individuals by addressing such issues as:

Power in Society. Women are barely tokens in the decision-making bodies of our nation. In Congress, women make up about 5% of the lawmakers. In state legislatures, the number is less than 20%. In order to achieve equal power in society, the gender movements seek to put more women into policy-making positions in government, business, education, religion, and all the other powerful institutions of society. Many men as well as women support feminist goals and encourage women to be politically active, to run for office from any political party, and to participate in the decision-making processes of the nation.

Traditional Roles. Often even feminists overlook the issue of homemaker's rights. There has historically been little recognition for the economic value of the vital services homemakers, whether male or female, perform for family and society. Those involved in the gender movements believe that legislation reflecting the reality of marriage as an equal economic partnership would go far to end discrimination against this group.

Because we all have heard and accepted that we live in a male-dominated society, it is easy to overlook some important and disturbing facts about what it could be like to be a modern male. While a small minority of men may indeed hold most of the power in our world, the majority of males are suffering from the deterioration of their traditional roles. The oppression of men can clearly be seen in alarming statistics on male addiction, suicide, and depression. Male teenagers are five times more likely to take their own lives than females. Overall, men commit suicide at four times the rate of women. Men between the ages of 18 and 29 suffer alcohol dependency at three times the rate of women of the same age group (see Figure 5.4). More than two-thirds of all alcoholics are men. Men account for more than 90 percent of arrests for alcohol and drug abuse violations. Sixty percent of all high school dropouts are males. Men's life expectancy is 10 percent shorter than women's (derived from Myers, 1995).

Figure 5.4
Gender oppression is seen in men's alcohol dependency, which is three times the rate of women's.

Both gender movements seek to broaden the definition of what it is to be male and female in our society. At the same time, there is clearly a need to consider the positive qualities of traditional roles and to explore emerging roles as well.

Economic Rights. Gender equity requires full economic equality between men and women. Eighty percent of Americans who are homeless are men. At the same time, female-headed families in the U.S. are four times as likely to be poor as male-headed or couple-headed families. The fight to end poverty focuses on equality in jobs, pay, credit, insurance, pensions, fringe benefits, and Social Security, through legislation, negotiation, labor organizing, education, and litigation.

Equal Rights Amendment. Though twice proposed in the U.S. Congress since 1923, the Equal Rights Amendment has yet to be ratified. Women are still not in the fundamental law of the land. Leaders of the gender movements believe that the ERA is essential to establish equality under the law for women, and the gender movements are committed to its passage as a priority. One right the ERA would guarantee is for women to serve in combat in the military and to perform traditional male tasks such as combat piloting. We have

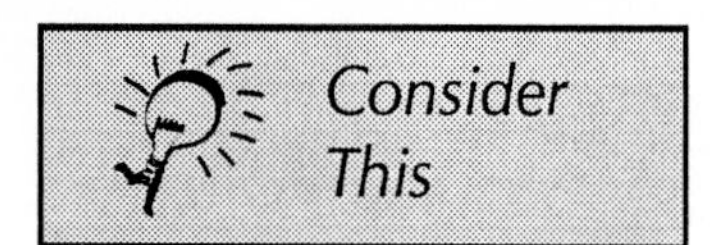

"Sugar and spice and all things nice,
And such are little girls made of.
Snips and snails and puppy dog tails,
And such are little boys made of."
—Robert Southey, poet

"A little girl is treated and molded differently from a little boy from the day she is born."
—Dr. Benjamin Spock, pediatrician and educator

been seeing, however, that many of the outcomes sought under the ERA are slowly being adopted without the passing of the amendment.

Reproductive Rights. The question of reproductive rights for women continues to fuel a controversy which deeply divides Americans of both genders. The so-called Pro-Choice advocates support access to safe and legal abortion and to effective birth control. They oppose attempts to restrict these rights through legislation, regulation, or Constitutional amendment. The so-called Pro-Life advocates oppose abortion on moral, ethical, and civil-rights grounds. They support alternatives such as adoption, abstinence, and pregnancy prevention. Beliefs about reproductive rights are frequently fraught with emotion and continue to result in violent confrontation. This is an issue that profoundly needs mutual understanding and one that transcends gender differences, encompassing personal philsophy and religion.

Custody and Child Support: Because our society has redefined gender roles and the meaning of "family," many people in the gender movements feel that laws concerning custody and child support need to be redefined accordingly. The gender movements seek public support for the vital role of fathers, stepfathers, foster fathers, uncles, brothers, and mentors in the growth and development of children while continuing support for the vital role of mothers and other female family members. The movements advocate joint custody for mothers and fathers, equality in child custody litigation, enforcement of children's rights of access to both parents, and equitable financial child support guidelines, orders, and enforcement.

Parental Leave Legislation. The so-called "Lost Father" syndrome refers to the fact that men too often do not take an active role in child rearing. To help remedy this situation, the gender movements support parental leave legislation, which gives working parents the right to take time from work to care for children. The movements also encourage fathers to push for changes in the workplace, including more flexible hours, part-time work, job sharing, and home-based employment so they can play more of a role in child care.

Harassment and Assault. More and more often, formerly feminist issues such as sexual harassment and gender discrimination in the workplace have become men's issues as well as women's. As our societal consciousness is raised, our justice system is recognizing that women aren't the only victims of sexual and physical abuse, including rape. Both women and men are frequently victims of sexual assault and spousal abuse. Rape and spousal assault legislation and support programs for battered women and men address the problem, but in spite of widespread attention and tremendous effort, the problem for both sexes continues to grow.

Education. In many communities–especially inner cities–men are often absent from the homes as well as from the schools. The gender movements have as a primary goal the re-establishment of a male voice in education. They seek to provide men with greater opportunities to be teachers and role models in early-grade classes. Such a male presence can help fatherless male children develop self-esteem.

Community Involvement. The gender movements seek to revive men's and women's concern for their community by encouraging the participation of men and women in community-based boys' and girls' clubs, scout troops, sports leagues, religious organizations, and big brother and sister programs.

Terms that belittle, demean, or stereotype men:

- hunk
- dork
- jock
- prick
- bastard
- pig
- effeminate
- mama's boy
- fag
- queer
- fruit
- fairy
- queen
- pansy
- wimp
- sissy

Terms that belittle, demean, or stereotype women:

- broad
- chick
- bimbo
- babe
- cow
- bitch
- butch
- dyke
- amazon
- honey
- lesbo
- tomboy
- the wife
- my old lady
- the little woman
- the ball and chain

(adapted from Papish, 1995)

Issues Regarding Sexual Orientation

- Harassment
- Inability to share relationships
- Transmittable diseases
- Self-awareness and esteem
- Disclosure or coming out
- Religious issues
- Finding peer support
- Fair and equal treatment
- Confidentiality
- Lack of legal support
- Finding positive role models
- Fitting some stereotypes
- Rejection from close friends
- Rejection from family
- Alienation from society
- Gay bashing and threats
- Fear of being alone
- Rejection from careers
- Sexism
- Acceptance and/or respect
- Living arrangements
- Pressures to get married
- Oppressive laws

(adapted from Papish, 1995)

Sexual Orientation

We all necessarily use labels to describe actions and attributes. However, it is problematic even for doctors and mental health professionals to find appropriate and accurate labels for sexual orientation. The subject is too complex, and human feelings, thoughts, and actions are seldom easy to objectify. We cannot use the term *sexual orientation* to divide people into two groups, heterosexual and homosexual. Rather, many people fall into some category between the two extremes. Sexuality is not black or white.

It is important to note that sexual behavior and sexual orientation are not the same. Some people engage in sexual activity with same sex partners who do not consider themselves gay or lesbian. There is no such thing as a homosexual or heterosexual *person*—there are only homosexual and heterosexual *activities* (see Figure 5.5). We use the terms *gay* (men), *lesbian* (women), or *homosexual* to describe people who participate in sexual activity with persons of their same sex. We use *heterosexual* or *straight* to describe those who participate in sexual activity with the opposite sex. *Bisexual* refers to men and women who may participate in sexual activity with both genders.

Figure 5.5
Sexual behavior and sexual orientation are not the same.

Our sexuality is an invisible quality. Scientists disagree about whether we choose our orientation, are led to it by our environment and social interactions, or are born with it. Many researchers now believe it is a combination of factors. Virtually all experts on human psychology agree that within everyone there exist both feminine and masculine traits. There is no "normal" or "abnormal" orientation, and homosexuality is not a disease (Reiss, 1980). Both aggression and sensitivity can be found in every emotionally healthy person, regardless of gender.

Incidence of Homosexuality

Suppose there are twenty people in your class, some male and some female. It's likely that two or more of your classmates are not heterosexual. Estimates of the incidence of homosexuality vary depending upon how strictly homosexuality is defined and the methodology of the study. The Kinsey Report of 1948 estimated that ten percent of the adult population was strictly homosexual, and most current estimates tend to lean toward the ten-percent level (see Figure 5.6). However, a University of Chicago survey in 1994, where researchers who were strangers to the respondents asked questions regarding sexual orientation, put the number at about three percent (Gallagher, 1994). Such low figures need to be considered carefully due to the methods utilized in the study. Due to social stigmas and people's reluctance to confide private information to a relative stranger, any statistic on sexuality must be taken with a grain of salt.

Your best friend may be gay or lesbian and you may not even know it. In maturity, people sometimes realize that they have a different sexual orientation than they believed themselves to have had in youth. As sexuality expert Margaret Hyde noted, many persons who are gay never admit their feelings to any of their friends. One cannot necessarily identify homosexual people by the way they talk, dress, or act (Hyde & Forsyth, 1994). Just as heterosexuals fall into many different types, so do homosexuals.

Figure 5.6
According to the Kinsey Scale, every person fits somewhere on the scale. Most people are somewhere in the middle. Remember that human feelings, thoughts, and actions are complex. Don't be quick to label yourself or other people based upon sexual activity or other outside factors.

attracted only to opposite gender (heterosexual)	occasional feelings toward same gender	equal attraction to both genders (bisexual)	occasional feelings toward opposite gender	attracted only to same gender (homosexual)

Figure 5.7
It is pointless to speculate about a person's private feelings, interests, instincts, emotions, and concerns.

Untrue Stereotypes of Gay Men:

- high voice
- gourmet cook
- fashion oriented
- one night stands
- hate women
- have A.I.D.S./H.I.V.
- limp wrists
- sissies
- small, slight build
- not interested in sports
- want to be women
- body builder
- physical weakness
- lisp
- recruit new gays
- effeminate
- molest children
- had distant fathers
- promiscuous
- good dancers
- had overbearing mothers
- female impersonators
- artistic

(adapted from Papish, 1995)

Labeling

Labeling not only promotes prejudice, but it is pointless when it refers to sexual orientation. It is impossible to know what someone feels or what his or her point of view is unless you are told by that person. We can observe behavior, but even behavior can be misleading. Many gay men and women are married to opposite sex partners, are parents, and in every way fit stereotypes for heterosexuals. Conversely, many heterosexual people fit stereotypes we may have for homosexuals. Women may dress in a masculine way, have short haircuts or deep voices, for example. Men may dress in flamboyant clothing, have long hair, or wear a lot of jewelry. We accept these variances from the traditional in many contexts every day. For instance, we don't label entertainers or people from different cultures as readily as we do those in our own social circles. If we look closely at our reactions and examine our labels, we will see how useless and sometimes silly they actually are.

Stereotyping

There are many different manifestations of sexual prejudice. *Chauvinism* is a belief that one gender is superior to the other. *Sexism* is discrimination against a particular gender. *Homophobia* is dislike or distrust stemming from feelings about a person's sexual orientation or lifestyle.

Biases and discrimination arise from and feed misinformation and fear. Some biases grow out of myths about sexual orientation. For instance, many people believe that men who feel a compulsion to wear women's clothes are homosexual. In fact, these men, called *transvestites,* are predominately heterosexual. *Pedophiles,* adult men or women who seek sexual satisfaction from children, fall into neither heterosexual or homosexual categories. Though pedophiles are often erroneously thought to be homosexual, it is not the gender of the child they are attracted to but rather the child's prepubescence. *Transsexuals* are individuals who have undergone or are in the process of undergoing a sex change operation because they do not like the gender with which they were born. Transsexualism is not related to sexual orientation, but rather has to do with physiology. A person who is *asexual* has no sexual feelings for either men or women.

Support Services

The National Organizations of Men and Women sponsor local support organizations for abused spouses, including shelters and counseling for abused spouses, gender support groups, and social groups. Gay-Lesbian-Bisexual owned and operated hotels, restaurants, bookstores, and travel agencies cater to the interests of their clients. Many churches welcome people of different sexual orientations and sponsor support groups. Telephone hotlines operate 24-hours a day in most cities, providing general counseling and crisis intervention for men and women of all orientations.

Interactions

Don't assume that anyone or everyone is heterosexual. Assume that you do not know anyone's sexual orientation for certain. Accordingly, do not use language that may be offensive.

Help to stop sexist behavior. Notice and ask people not to tell sexist jokes or use sexist language. At the same time, try to acknowledge when someone takes even small steps in the direction of tolerance and sensitivity.

Avoid discussing others' behavior. Refer only to your own behavior.

Ask a Friend

Have you ever seen a person on the street and thought he or she was gay or lesbian? What dress, features, or behavior made you think so? What other factors besides sexual orientation may have explained what you saw? How valuable was your labeling?

Untrue Stereotypes of Lesbian Women:

- masculine
- short hair
- wear men's clothes
- aggressive
- tough
- non-sexual
- fat/husky
- short
- want to be men
- athletic
- hiking boots
- wear flannel shirts
- strong
- hairy legs
- have tatoos
- ugly
- not attractive

(adapted from Papish, 1995)

Before I turned 14, I knew I was gay. It wasn't an easy realization because I grew up hearing negative comments and jokes about homosexuality. I was very afraid that my parents, teachers, and classmates would change their impressions of me if they ever discovered my sexual orientation. I even developed several health problems as a result of the turmoil that I felt.

One night I vented my frustrations in a letter to my parents. I never intended them to read it, at least not consciously. My mother found the note the next day while I was away at school, and she left another note for me in its place. She was deeply saddened by the entire situation: my sexuality in general, the stress I had forced upon me, and my fear of discussing it with anyone. My parents then sent me to a therapist to discuss my emotions and to attempt to "repair the problem."

After lots of therapy, I ended up feeling even better about my sexuality than before. Family counseling later encouraged my parents to love me for who I was. I then had the self-esteem to come out to my friends and relatives. I have since felt very positive about my life.
—Jeffery B., student

Work to safeguard the basic rights of all people. Some people may be uncomfortable with their own sexuality and may need friendly support.

Be sensitive to labels. Call people what they prefer to be called, and make an effort to determine what is appropriate in a given situation. If you are unsure what designations to use, ask. But avoid labeling people whenever possible. Differences should be mentioned only when relevant.

Avoid biased speech and writing. We can help to remove bias in our speech and in writing by using terms such as "same-gender," "male-male," "female-female," and "male-female" sexual behavior when referring to sexual pairings.

Avoid patronizing. Sometimes when we try to stretch beyond our prejudices, we make patronizing statements unintentionally, such as "You don't *seem* gay."

Treat others as you would like to be treated. Sexual orientation has nothing to do with ordinary daily living. Treating others as you would like to be treated yourself is just a simple matter of respect for human diversity.

Demonstrate change. If people are responding negatively to you, instead of sticking on that point, show them something positive to which they can respond. Draw upon their optimism rather than their fears. Find a common point (for example, you both cheer for the same team) so that it gives the person something different to focus upon. We can break down prejudices one person at a time. Provide a model through your own behavior. You will experience change in yourself as you change other people (Kaye, 1994).

Terminology

The following additional terms relating to gender and sexual orientation are also important to consider.

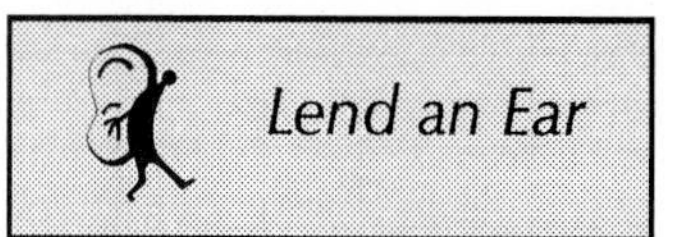

"A human being is a multisexual being and contains the possibility of all sexual expressions simultaneously within itself. In any moment, we can express any of them. If we relinquish dualistic judgments, we are simply whatever we are."
—Master Charles, spiritual leader

Androgynous. This refers to a person who exhibits masculine and feminine qualities, appearances, or behaviors.

Asexualism. A person who is asexual has no sexual feelings for either men or women.

Celibacy. This is a choice not to participate in sexual activity. Some people practice celibacy in order to avoid sexually transmitted diseases. Others practice celibacy until they find a life partner. Celibacy affects one's sexual activity—it does not change one's sexual identity.

Coming out. This refers to the process of accepting one's own sexual orientation and telling others about it (see Figure 5.8). This process is often difficult because it involves both personal and social consequences.

Figure 5.8
"Coming out," short for "coming out of the closet," refers to being open about one's sexual orientation.

Heterosexism. This refers to discrimination based upon the assumption that heterosexuality is the only viable or acceptable orientation.

Homophobia. This is a dislike or distrust stemming from feelings about a person's sexual orientation or lifestyle.

Sexual harassment. This includes, but is not limited to, name-calling, innuendo, insults, obscene gestures, gay-bashing jokes, physical threats, and actual physical violence.

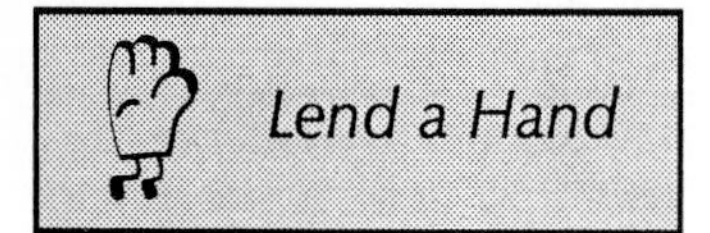

Intervene whenever you notice that someone is being constrained by gender stereotypes, or criticized by others for acting on one's own interests and feelings.

Conclusion

The varieties of human sexuality defy easy categorization, even though labels are pervasive in our society. Though a minority of people are strictly heterosexual and strictly homosexual, most others fall somewhere in the middle. Biases about gender, sexual orientation, and men's and women's "natural roles" have led to sexual discrimination. Until we remove sexism from our institutions and sexist attitudes from ourselves, human beings will continue to be limited (Hanmer, 1990).

References

Gallagher, J. (1994). 10%: reality or myth? The Advocate 15 Nov., 23.

Hanmer, T. J. (1990). *Taking a stand against sexism and sex discrimination.* New York: Franklin Watts.

Hyde, M. O. & Forsyth, E. H. (1994). *Know about gays and lesbians.* Brookfield, CT: Millbrook Press.

Kaye, K. (1994). *Workplace wars and how to end them.* New York: American Management Association.

Myers, D. G. (1995). *Psychology.* New York: Hope.

Papish, R. (1995). Diverse works. Gainesville, FL: University of Florida.

Reiss, B. F. (1980). Psychological tests in homosexuality. In J. Marmor, (Ed.), *Homosexual behavior* (pp. 296-311). New York: Basic Books.

Schuman, D. & Olufs, D. (1995). *Diversity on campus.* Boston: Allyn and Bacon.

Suggested Readings

Berbards, N. (1989). *Male/female roles: Opposing viewpoints.* San Diego: Greenhaven Press.

In this book, various authors debate how sex roles were established and how men and women respond to changes in sex roles.

Deegan, M. J. & Brooks, N. A. (1985). *Women and disability: A double handicap.* New York: Transaction Books.

This book discusses the way disability magnifies the effects of sexism on women and the way that disability changes relations of disabled women with men and with other women.

McCauslin, M. (1992). *The facts about lesbian and gay rights.* New York: Crestwood House.

This book examines various myths, fears, and misconceptions about homosexuality and current attempts to gain fair treatment of homosexuals in the areas of housing, the media, the church, and the military.

Persing, B. S. (1983). *The nonsexist communicator: Solving the problems of gender and awkwardness in modern English.* Englewood Cliffs, NJ: Prentice-Hall.

This book proposes ways to avoid using sexist language and suggests non-gender-specific alternatives.

Notes

Individual Activities

1. As you get dressed in the morning, observe what articles of clothing you put on. What elements of your wardrobe or personal grooming adhere to a gender stereotype? Write your thoughts below:

2. Think back on your own childhood and on children whom you know today. In our culture, how are children led to stereotypical gender-related interests through toys, books, games, etc.? Write your observations below:

Group | Activities

Group Activities:

1. As a group, think of role models who transcend stereotypes. Write the names of the role models below and explain why you chose them:

2. Imagine that you are presented with the business cards below. What is your immediate assumption about their owners' gender and/or sexual orientation?

Dr. N. R. Jones	**Pat Smith**	**Chris Brown**	**Capt. Terry L. Black**
Dept. of Women's Studies	*Smith's Gym*	*Massage Therapy*	*U.S. Air Force*

Discuss with your group why you made certain initial assumptions about these people. Write your reactions below:

Reflection Paper 5.1

Do you think it is beneficial for persons who are gay or lesbian to come out? Why or why not?

Reflection Paper 5.2

Whether intentionally or not, our culture practices sexual discrimination all the time. Name three examples of discrimination that you have witnessed and propose solutions.

Notes

Chapter 6 Religion and Belief Systems

Chapter Sections

"Tolerance implies no lack of commitment to one's own beliefs. Rather, it condemns the oppression or persecution of others."
—John F. Kennedy,
President of the U.S.

Objectives

- Explain the concept of religion.
- Identify the major and minor belief systems of the world.
- Outline the necessary steps for fostering an understanding of people with different beliefs.
- Describe how once can view his or her personal religion in a universal context.

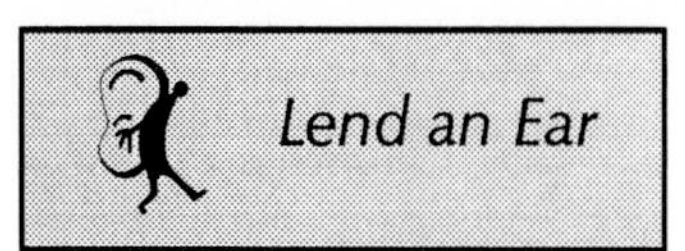

"In Man, the positive content of religion is the instinctive sense—whether conscious or subconscious—of an inner unity and continuity with the world around. This is the stuff out of which religion is made." —Edward Carpenter, philosopher

Introduction

Religion involves powerfully-charged feelings, passionate dedications, and deep loyalties. Religion is as old as humankind. In a very real sense, the history of humankind has been driven by religion. In the name of religion, wars have been fought, new territories have been discovered, and conflicts have been resolved. Though there are myriad expressions of spirituality across the globe, the underlying religious impulse is shared by all faiths (see Figure 6.1). The purpose of this chapter is to define what religion means, present an overview of the major and minor belief systems of the world, and examine steps you can take to understand people who follow faiths different from your own. The ultimate aim of this chapter is for you to view your own religion in a universal context.

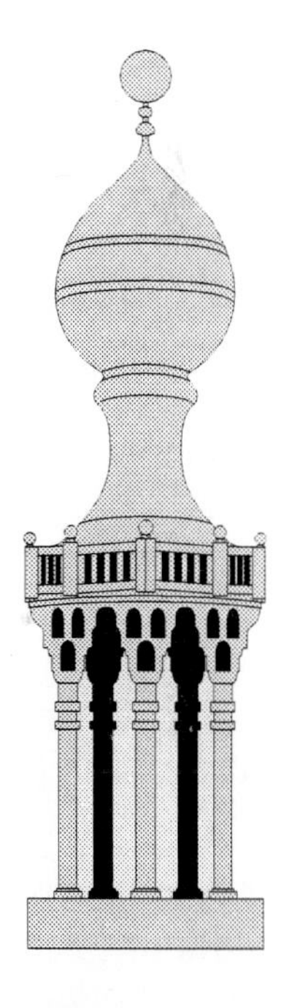

Figure 6.1
The Muslim prayer tower is one expression of the religious impulse.

The Culture of Religion

Our culture is enriched by religious traditions. The passing of our years is marked by festivals, rituals, and holidays which honor figures and events from many faiths. Such celebrations typically involve special

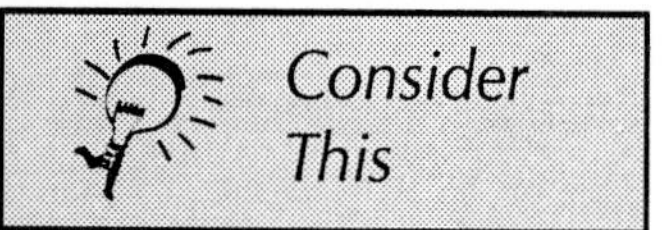

Ancestor worship is a facet of many Eastern religions. Altars are set up in a family house in reverence of deceased family members. Here, on the birthdays of the deceased, candles are lit, prayers are said, and food is offered. At the grave site, monuments and arches are erected. In what similar ways does your own culture and religion honor its dead?

costumes, foods, dances, and songs. The constant repetition of religious stories has kept alive some of the world's finest literature through the ages. Much of the most beautiful architecture and art in the world, from the pyramids in Egypt to the Sistine Chapel in Rome, have come from our attempt to express religious feelings.

Not all religions survive. The gods of Ancient Greece, for instance, are no longer worshipped. It takes people to keep a religion alive, and people incorporate religion into their daily lives in such profound ways that we are often unaware of them. People of many faiths welcome the coming of spring with a religious-based holiday, the harvest with a thanksgiving feast, and the new year with a celebration often related to religious beliefs. Even those who profess no faith usually participate in rituals such as weddings, funerals, and baby-showers, all of which have their roots in religious rituals.

What is Religion?

Religion has been defined as the attitude of individuals in a community toward the powers which they conceive as having ultimate control over their destinies and interests (Lewis, 1968). But don't let that definition confuse you. Religion is simply the part of culture associated with people's deepest convictions (see Figure 6.2). Through religion we seek to explain awe-inspiring, mysterious concepts: What is God? Why does evil exist? Is there a soul? What happens when we die? How did humanity begin and when, if ever, will it end? People need general ideas that give meaning to their lives by explaining their place in the universe. Different religions attempt to give different answers to these human questions.

Figure 6.2
Religion is associated with deeply personal beliefs about the meaning of life.

There is talk today that the world is becoming one and that people are uniting under a "New World

Order." Innovations in technology are certainly tying people together. Over the worldwide computer Internet, for example, a Baptist student in Kansas can bump into a Buddhist student in Japan and strike up a conversation. Although we live in a "global village," not everyone agrees about how to live, how to govern, or how to worship. Yet if we intend to live together, we must understand each other's deepest convictions. The majority of the world's population adheres to one of three religious categories: Hindu, Buddhist, or Judeo-Christian-Islamic. In order to appreciate and understand another belief system, we do not necessarily need to agree with it. Rather, we need only seek to understand and respect the differing viewpoint and the person who holds it.

The Major Religions of the World

It is impossible in this context to fully discuss every religion practiced in the world, much less the particular denominations and offshoots within each religion (see Figure 6.3). However, a brief overview of the world's major belief systems is practical and useful. The following summaries, therefore, are intended to introduce some key concepts at the foundation of each religion.

Figure 6.3

World Membership of Major Religions

Christians	1,927,953,000
Moslems	1,099,634,000
Nonreligious	924,078,000
Hindus	780,547,000
Buddhists	323,894,000
Atheists	219,925,000
Taoists, Chinese folk beliefs	225,137,000
New Religionists	121,297,000
Tribal, Shamanists	111,777,000
Sikhs	19,161,000
Jews	14,117,000
Baha'is	6,104,000
Confucians	5,254,000
Jains	4,886,000
Shintoists	2,844,000
Other religionists	70,183,000

Caution: Definitions of membership vary greatly from one religious body to another. For example, some count children who have been received into the faith, and others count only adults (Famighetti, 1996).

Animism/Shamanism

Animism and Shamanism are religious systems typically found in the tribal societies of Africa, Australia, and North and South America. Animism is a belief that God is present everywhere, in a multiplicity of expressions that inhabit natural objects such as rocks and trees and rivers. All beings and things were

Imagine that you belong to a particular religious minority. What day-to-day experiences or encounters might be striking reminders that the people around you don't share your faith?

created by God and can be used for either good or bad purposes. Gifted people called Shamen (priests, prophets, medicine men) believe they can go into a visionary state and communicate with the great power of the universe. Such a communion helps to sustain nature and the relationship between human beings and the universe.

Buddhism

Buddhists are found in the greatest numbers in eastern Asia. Buddhism is a religion founded by Buddha, a prince who lived in India several hundred years before Jesus. This prince renounced his wealth and status and began teaching a philosophy of physical discipline, moderation, silent contemplation (meditation), and universal brotherhood as a means of liberation from the physical world. Buddhists believe that our desires trap us in the "Wheel of Life," a cycle of rebirths. The goal of the Buddhist is to attain Nirvana, a state of complete peace and bliss in which one is free from the inability to fulfill desire.

Christianity

Christianity has its roots in Judaism. Christianity is a religion of love, compassion, and fellowship, based on the life and teachings of Jesus Christ. Christianity sees God as a trinity—the mystery of three in one. Christians believe that Jesus is the Messiah (or savior) sent by God. They believe that Jesus, by dying and being resurrected, made up for the sin of Adam and thus redeemed the world, allowing all who believe in Him to enter Heaven. Christians rely on the *Bible* as the inspired word of God.

Confucianism

Confucianism is a system of ethics based upon the teachings of Confucius. This system has dominated Chinese culture for 2,000 years. Confucius emphasized moral perfection for the individual and social order for society. He believed world harmony could be attained if people possessed such virtues as loyalty, respect, integrity, piety, righteousness, wisdom, benevolence, and courage. While Confucianism is not strictly a religion, its followers are certainly religious. Confucianism has no priesthood, churches, or system of gods, but Confucius himself is revered and worshipped as the ideal man, and Confucians believe that the goodness in human nature comes from Heaven.

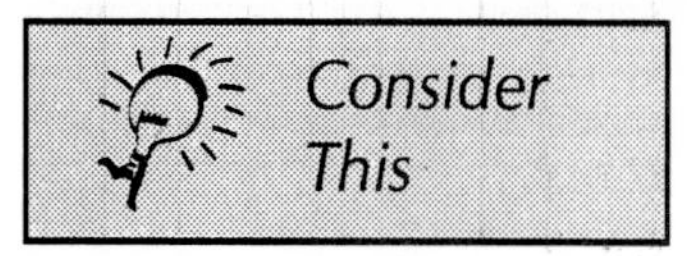

Do you believe that all of your actions, both good and bad, have consequences? If you do, then you understand Karma, the law of cause and effect.

Hinduism

Hinduism, a religion of India, is one of the world's oldest living religions. Its holy scriptures, the *Rig-Veda*, date all the way back to 4000 B.C. and make reference to the genesis of the universe. Hindus believe in a personal creator God who sustains the universe. They follow the principle of Karma, a law of cause and effect in which every action one does, whether good or bad, eventually comes back to him or her. Hindus also believe in reincarnation, a cycle of never-ending births and deaths in which ignorant people are reborn according to the deeds (or Karma) of their past lives. Enlightened people are not reborn. Hindus see the material world as being an illusion which can be changed by one's viewpoint. They also believe that all people are actually facets of Brahman, the eternal web of the universe.

Untrue Stereotypes of Persons of Different Faiths

- They are ignorant or unenlightened
- They are misguided
- They are sinful,"ungodly," or devil-worshippers
- They are strange
- They are fanatics
- They want to convert you to their religion
- Their religious heritage is primitive
- All are religious
- All are experts on their religion
- All observe holidays or religious practices
- All hate people who don't adhere to their faith

Islam

Islam is the dominant faith in the Arab nations and is growing in other parts of the world. Like Christianity, Islam has its roots in Judaism and worships the same indivisible God (called Allah in Arabic). Islam was founded by Mohammed, a prophet (messenger) of God who dictated the *Koran*. The fundamental belief of Islam is that "There is only one God, and Mohammed is his prophet." The word *Islam* means "submission to the will of God." Followers of Mohammed, called Moslems, are obliged to pray five times a day, to avoid pork and alcohol, to give to the poor, and to make a pilgrimage to

Mecca (Mohammed's birthplace) at least once in their lives.

Jainism

Jainism is a native religion of India, founded by a man named Mahavira. This religion grew out of Hinduism and teaches a doctrine of non-injury, or *ahimsa*. Jains believe in the law of Karma (cause and effect) and reincarnation. Jains do not eat meat, and they seek to avoid harming anything believed to have a soul. Jains attempt to achieve *moksha,* or salvation, through self-discipline, knowledge, faith, and right conduct.

Judaism

Judaism, the religion of the Hebrews, centers around a personal God, Yahweh. A succession of great prophets (spiritual messengers), such as Abraham, Moses, Elijah, and Isaiah, spoke in God's name and taught Jews how to know and serve Yahweh. In Judaism, a permanent covenant with God and his chosen people allows heaven to come to earth. Jewish law, legend, and history is embodied in the *Talmud* and the *Torah.* Jewish worship takes place in a synagogue and is led by a rabbi (a scholar of Jewish thought). Judaism is the foundation of the Judeo-Christian-Islamic tradition.

Shinto

Shinto is the national religion of Japan. The word *Shinto* means "The Way of the Gods." Many Gods are honored, representing natural forces, ancestors, former emperors, and national heroes. Over 100,000 shrines have been built to the Gods. However, these shrines are not meant for the assembling of worshippers but rather as dwellings for the Gods. Many of the shrines are very small and are focal points for festivals and patriotic holidays. Shinto promotes national loyalty.

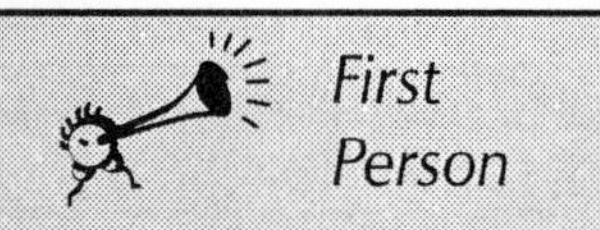

First Person

I always thought talk about religious persecution was just paranoia until last year, when the son of a friend of mine was chased by a gang of his sixth grade classmates and beaten up. The boy is Jewish and some boys in his class started playing a game called "Nazis." They had all been friends before, but now they targeted him to harass for fun.

The teacher sat them all down and told them about what Nazism is about and explained what Judaism is. They discussed the importance of religious freedom and so on, and the games stopped, at least at school.

I know that my little friend will never take his safety for granted again, though, and neither will I. There is so much fear of anything different from the majority, and it is so near the surface. The scariest thing is that wherever you go in the world, somebody is different.

—Anna M., parent

Sikhism

Sikhism is a faith that arose in India as a result of the coming of Islam. It features elements of Islam and Hinduism, but maintains a separate identity. The founder of the movement was named Nanak (see Figure 6.4). He believed in the equality of humankind and taught that the way of salvation was through *bhakti* (devotion). Sikhs worship Hari, the one God. Sikhs believe in reincarnation and hope to break out of an endless cycle of rebirths to merge with the soul of God. The sacred book of the Sikhs, the *Granth,* contains Nanak's poems and songs of devotion.

Figure 6.4
Guru Nanak, founder of Sikhism, taught that all people were equal.

Taoism and Chinese Folk Religion

Taoism is a philosophical religion native to China. It was founded by a great philosopher named Lao Tse. Taoists attempt to live according to the Tao (pronounced *DOW)* , or "Way," which they believe governs the universe. Lao Tse called for people to be peaceful, inactive, and quiet. He believed that bringing these qualities into daily life would put one effortlessly in touch with the universe. The *Tao Te Ching,* or Book of the Way, spells out the doctrine of Lao Tse. Chinese folk religions combine Taoism, Buddhism, Confucianism, ancestor worship, and local deities.

New Religions

New Religions have grown up in Asia, mostly since 1945. Some of these popular new beliefs were inspired by Buddhism and Shintoism. They were founded by visionaries who preached a fresh new social ethic. Most stress gratitude for God's creation, social rather than individual good, and hard manual labor. The Nichiren Shosu religion, founded in Japan earlier this century, has spread to other countries in both the East and the West.

Issues Persons of Different Faiths May Face

- Understanding the cultural issues of their religion
- Dating and marriage within or outside the faith
- Religious persecution
- Stereotypes about people of their religion
- Other people being ignorant about their religious culture
- Group identification
- Understanding the importance or lack of importance of certain holidays
- Finding peer support
- Not knowing a lot about their culture themselves
- Understanding religious practices
- Finding mentors or role models
- Maintaining a cultural identity
- Careers that allow for cultural observance and participation
- Maintaining self-esteem and pride
- Comfort wearing special clothing or other symbols of religious pride and observance
- Stigma of being identified as a minority

Denominations and Minor Religions

Within the major religious systems are numerous denominations, all with their own special beliefs and practices. For example, Christianity is divided into four main denominations: Catholic, Protestant, Orthodox, and Anglican. Each of those is divided into smaller affiliations—Methodist, Baptist, Lutheran, Pentecostal, Presbyterian, Seventh-Day Adventist, Jehovah's Witness, Greek Orthodox, Shaker, and so on.

In addition to the major religions, there are dozens of minor religions, branches, and offshoots, including:

Celtic religions. These include the ancient nature cults of Great Britain, such as Druidism.

Caribbean religions. These religions of the Caribbean islands include Voodoo and Santeria. Though frequently misunderstood and stereotyped, Caribbean religions have a unique and rich cultural heritage. Many combine elements of Catholicism and traditional belief systems.

Eckanar. Eckanar focuses on the personal experience of the sound and light of God.

Esoteric Brotherhoods. These philosophical societies include the Rosicrucians and the Freemasons. They have a long history, dating back to ancient times. Combining science and religion, they encourage right living through strength of character, self-reliance, justice, unflinching courage, honesty, logic, and compassion.

Kardecism. With many followers in Brazil, Kardecism originated in France in 1854. Founded by Allan Kardec, the religion is similar to Buddhism in that it seeks human evolution to a level of supreme goodness and peace, through as many reincarnations as necessary to purify the soul. Unlike Buddhism, Kardecism recognizes Jesus Christ as the messiah.

Krishna Consciousness. Originating in India, this movement was made popular in the West during the 1960s by A. C. Bhaktivedanta Swami Prabhupada. It focuses on love, service, and devotion to Krishna, the personal aspect of God. Followers of Krishna avoid eating meat, engaging in illicit sex, the use of intoxicants, and gambling. They frequently chant the name of God (Prabhupada, 1969).

Latter-Day Saints. The Mormon church originated in the United States in 1830. Its teachings are based upon the Bible and the *Book of Mormon,* which was revealed to church founder Joseph Smith and is a record of the early inhabitants of America. Mormonism stresses abstinence from liquor, tobacco, tea and coffee. The church encourages hard work and a strong family life.

Magick systems. The ancient practice of Witchcraft is a magick system. Though often confused with Satanism, many magick systems practice "white magic" and seek personal empowerment through the energy of nature.

Natural Law systems. These are not religions in terms of doctrines or dogmas. Practitioners of these systems view individual life from a universal perspective. Based upon modern and ancient science and reports of higher states of consciousness, they recognize the intelligence and creativity underlying the order in the universe. Such a recognition offers one the choice regarding a personal relationship with the divine. Practitioners observe the natural rhythms and cycles of the universe, including the cycle of creation-maintenance-evolution-dissolution.

North, Central, and South American Indigenous Religions. These are ancient regional belief systems, many of which are tribal in origin. Followers of these religions are usually descended from a particular area's

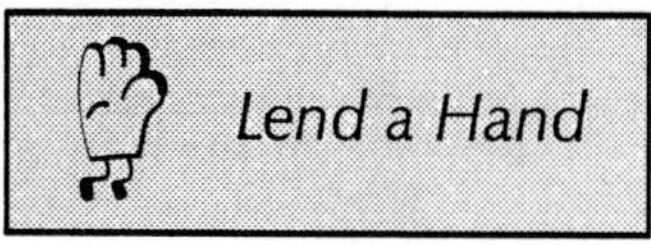

Does your belief system teach that people should help their fellow humans in times of need? Why not practice what you preach and volunteer your time and abilities to a good cause?

native population.

Scientology. This movement, founded by L. Ron Hubbard, teaches a modern science of mental health and social betterment. Scientology offers principles for improving self confidence, intelligence, and ability.

Spiritualist systems. These systems include New Age beliefs (such as channeling).

Sufism. This is a system of Islamic mysticism which teaches that repentance, abstinence, poverty, patience, and trust lead to union with God. Love is the key to Sufi ethics. Some Sufis, called dervishes, engage in a devotional exercise which involves a whirling dance.

Transcendental Meditation. The Transcendental Meditation movement, developed by Maharishi Mahesh Yogi, is a natural law system which can be followed from within any religious tradition (Yogi, 1972). This movement encourages people to spend a few minutes sitting in silence, twice a day (see Figure 6.5). Meditation is said to reduce stress, improve health, and foster spiritual growth.

Unification Church. The Unification Church was established in Korea by the Rev. Sun Myung Moon in 1954. The church combines wisdom from numerous world religions as it works toward bringing about spiritual and social reform on a global scale.

Zoroastrianism. This is a religion founded in Iran, based upon the philosophy of a man named Zoroaster. Zoroastrians worship the god Ahura Mazda, creator of goodness and light, and view life as a constant battle between good and evil. Good thoughts and deeds are the keys to salvation.

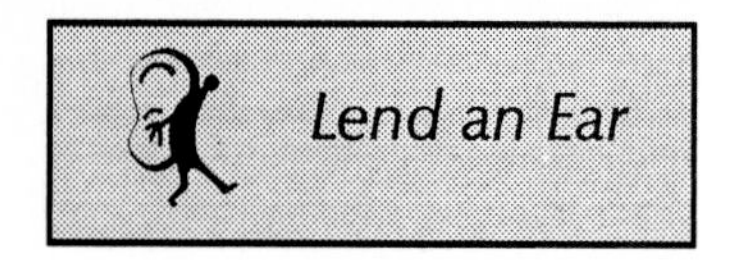

"Evolution means respect. We all have to learn that. Respect, fellow friends, respect!"
—Divaldo Franco, Kardecism leader

Figure 6.5
Meditation, a practice of sitting quietly for a period each day, is encouraged by numerous religious traditions.

Other Categories of Belief

There are, of course, people who do not adhere to any faith at all. Some people call themselves skeptics or Agnostics. Agnosticism is a denial of knowledge about whether or not there is a God. People who are agnostic are open to the possibility that God exists but believe that there can be no proof either way. Other people, called Atheists, go so far as to reject all religions outright. Atheism is the denial that there is any God, no matter how God is defined.

On the opposite end of the spectrum, there are people who follow the wisdom of more than one faith. The Unitarian and Universalist movements, for example, study both Eastern and Western spiritual writings in a quest for universal brotherhood. Similarly, the Baha'i faith (which originated in Iran) emphasizes the spiritual unity of all humankind. A movement called Ecumenism promotes global unity among religions through greater cooperation and improved understanding. Another specialized church is the Metropolitan Community Church, an international Christian organization that welcomes persons who are lesbian, gay, or bisexual, and their friends and family.

Figure 6.6
All religions have the same roots in deep convictions. All are held up by a common search for answers. From that trunk, different belief systems branch out. The leaves represent different offshoots, each with its own special practices. (Figure derived from Conley, 1997.)

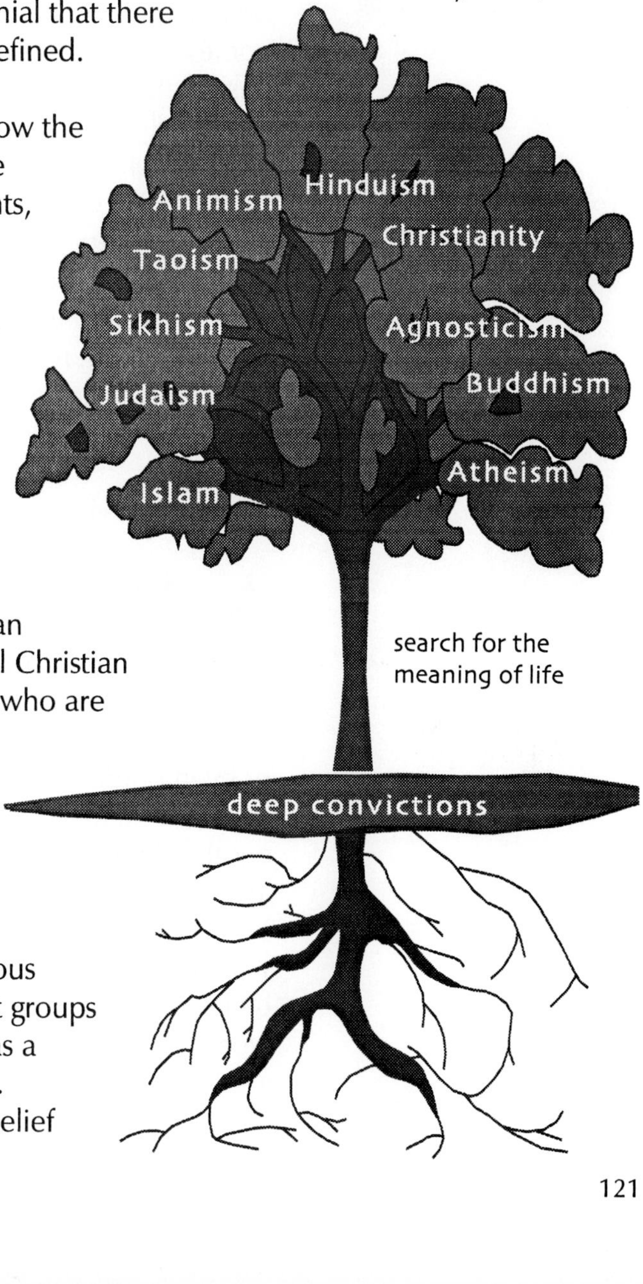

Religious Prejudice

Unlike racial prejudice, religious prejudice is directed against groups that people *choose* to join as a matter of faith (Kronenwetter, 1993). Religious prejudice stems from the belief

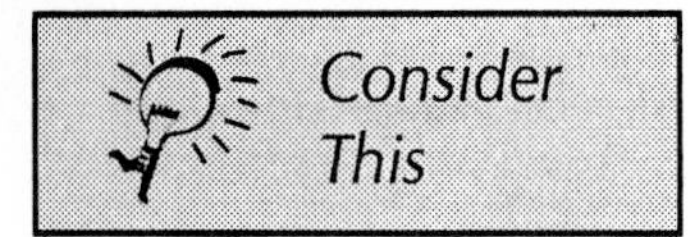

When discussing diverse belief systems, build upon what people already know. Always start with familiar terminology, and highlight similarities before pointing out differences.

that one's particular faith is favored by God. Ironically, virtually all of the world's religions teach against prejudice in favor of tolerance, unity, and love.

Perhaps people will never agree about the fine points of religious doctrine, but such agreement is not necessary. One does not have to convert to another religion in order to honor the piety of its followers (see Figure 6.6). The challenge is to practice what you preach. Do you believe that your actions will one day be judged by God, or that your deeds in this life will come back to you the next time around, or that you are answerable only to yourself? No matter how we express it, virtually everyone agrees that people are all accountable for their actions. Whether we are facets of God, or part of the web of the universe, or descendants of Adam, everyone is entitled to respect. Regardless of our religion or lack thereof, we are all travelers on the journey of life. If you are convinced that yours is the best path, be a respectable representative and give others the opportunity to be drawn to your example.

Support Services

Most religions have formal places of worship where members of the particular faith can gather together. Such places of worship are listed in the Yellow Pages. Some smaller congregations may meet in houses, in rooms rented by other churches, or even outdoors. Wherever they meet, members of a particular faith form social networks to support one another in times of crisis or need and to sponsor activities for singles, children, adolescents, families, and senior citizens. Many churches offer evening classes which provide additional instruction on their particular belief system or on other faiths practiced around the world.

Terminology

In addition to the major and minor world religions defined in this chapter, it is useful to be familiar with these other related terms:

Belief. This refers to faith and trust in the truth or existence of something abstract or intangible.

Conviction. This is an earnest and profound belief in something, such as religious doctrines.

Creed. A creed is an accepted system of religious belief. The word is derived from the Latin *credo,* meaning "I believe."

Doctrine. This is a particular principle taught by a religion.

Dogma. This refers to a set of firmly established doctrines authoritatively put forth by a particular church.

Faith. This refers to a belief in God or the teachings of a particular religion. This word is also used to mean a system of religious belief, as in "the Jewish faith."

Persecution. This refers to the act of oppressing, injuring, subjugating, and/or exterminating someone for adhering to a particular religious faith.

Principle. This is a basic law or rule at the foundation of a belief system..

Tenet. This is a doctrine that a person or religion holds as true.

Who is a famous leader you admire who was devout in a particular faith? Examples include Martin Luther King, Jr., Mother Teresa, Mahatma Gandhi, Malcom X, Billy Graham, and the Dalai Lama.

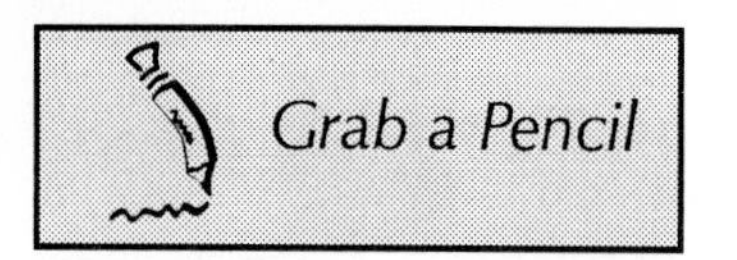

Jot down five things you would like to know about a different religion. Where can you go for answers?

Interactions

Resist trying to force others with another faith into the categories of your own belief system. Take them on their own terms, giving their experiences and beliefs respect.

If another religion seems foreign or confusing, remember that it is simply another human attempt to make sense of life's mysteries. Keep in mind that *your* belief system may seem just as foreign to someone who doesn't follow it.

Before you form an opinion about a believer of another faith, remember the key quality you have in common. Both of you carry on the tradition of a long line of believers stretching back through time.

Do not be embarrassed by your ignorance of a person's faith and customs. Ask questions in an open, friendly, interested manner and you are likely to find that the person will be eager to discuss and share his or her religious culture.

Approach another religion in the light of what it contributes to the life of its followers. Ask followers of this religion how their faith has changed their lives, or empowered them, or comforted them. Ask how religious holidays involve the participation of their families.

Conduct frank discussions about religious persecution. Consider watching with a friend one of several compelling films about Anne Frank, the young Jewish girl who wrote her famous diary while in hiding from the Nazis.

Conclusion

Religion is a universal experience of awe and wonder in the presence of the mysterious. In different lands and different times, throughout human history, religion has helped people to understand their lives in a larger, cosmic context. Religion concerns the deepest and most sensitive area of a person's experience. The only way to be sympathetic and objective toward others of another faith is to be knowledgeable about their belief system.

References

Conley, C. (1997). *On top of the world: A life planner for every student.* Manuscript in publication.

Famighetti, R. (1996). *The world almanac and book of facts 1997.* Mahwah, NJ: KIII Reference Corporation.

Kronenwetter, M. (1993). *Prejudice in america: Causes and cures.* New York: Franklin Watts.

Lewis, J. (1968). *Religions of the world made simple.* Garden City, NY: Doubleday.

Prabhupada, A. C. B. S. ((1969). *Sri Isopanisad.* Los Angeles: Bhaktivendanta Book Trust.

Yogi, M. M. (1972). *The science of creative intelligence: Teacher training course.* Fairfield, IA: Maharishi International University.

Suggested Readings

Carmody, D. & Carmody, J. (1983). *Eastern ways to the center.* Belmont, CA: Wadsworth Publishing Co.

This is a readable introduction to the belief systems of Asia.

Eliade, M. (1987). *The encyclopedia of religion.* New York: Macmillan.
This encyclopedia breaks down world religions into geographical categories and features informative, comprehensive discussions of each faith and its offshoots.

Ferm, V. (1945). *Encyclopedia of religion.* New York: Philosophical Library.
This encyclopedia offers essay-length discussions of world religions and important figures.

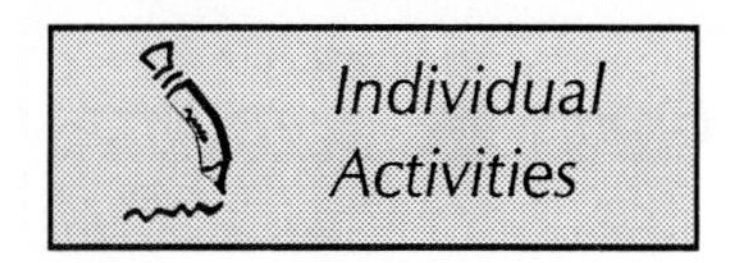

1. Compare the wisdom of another scripture (such as the *Koran, Rig-Veda,* or *Talmud)* to the wisdom of your own scriptures.

What similarities do you notice?

What differences do you notice?

2. Visit the church of another faith. You will notice that every religion has a system of rituals or ceremonies, no matter how elaborate or simple. Even a bowed head, folded hands, and the repetition of sacred words are ceremonial. Describe a ceremony you witness. How is it similar to or different from a ceremony with which you are more familiar?

1. Have each person in your group represent a different faith. Defend your assigned faith against charges that it is "strange," "illogical," "ungodly," etc.

Reactions:

2. Have each person in your group explain his or her particular belief system. Why do you believe what you do? What does your belief system teach about human diversity?

Reactions:

Reflection Paper 6.1

Most world religions teach that we are all part of a larger design. Discuss some ways in which an understanding of this common point could lead to universal brotherhood.

Reflection Paper 6.2

Reflection Paper 6.2

Even Atheists are profoundly affected by the world's religions. Discuss how religious belief affects everyone's lives every day.

Notes

Chapter 7 Socioeconomic Diversity

Chapter Sections

"When I was young I used to think that money is the most important thing in life. Now that I'm older I know that it is."
–Oscar Wilde

Objectives

- Expain how society determines socioeconomic status.
- Describe the majority of residents of designated "poverty areas."
- Identify the ways of measuring quality of life.
- Identify the best methods of overcoming socioeconomic stereotypes.

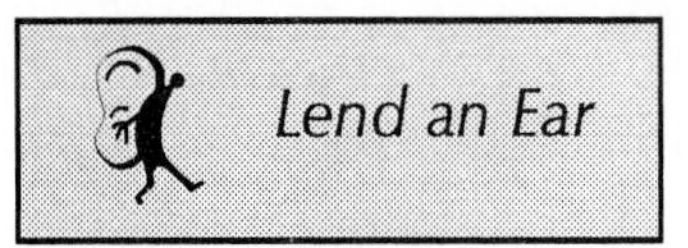

"Unfortunately, many Americans live on the outskirts of hope—some because of their poverty, some because of their color, and all too many because of both. Our task is to help replace their despair with opportunity." —Lyndon Johnson, President

Introduction

We have looked at diversity in race, ethnicity, gender, sexual orientation, and religion and have seen how stereotypes and prejudices affect nearly every person in one way or another. Now let's examine socioeconomic diversity. Its importance is clear from its name. We add the prefix "socio" because economics affects our status, or place in society, and therefore colors every aspect of our lives. The level of our wealth or poverty can lessen or increase our chances of overcoming other differences. Wealth gives us access to services, advantages, and opportunities which are unavailable to those with economic disadvantages, and it can break down social barriers because people often put money before other biases (see Figure 7.1).

The term "socioeconomic status" describes the place in a society where a person's income, level of education, or occupation places him or her. The United States Bureau of the Census is responsible for measuring economic conditions using these criteria. In this chapter we will look at data on the socioeconomic status of Americans and consider how wealth and poverty are related to some societal issues which affect us all. We

Figure 7.1
Wealth opens doors to opportunities and advantages, breaking down social barriers.

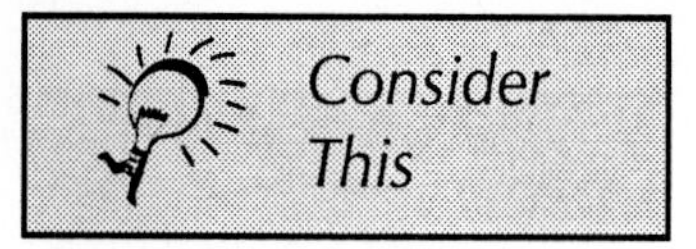

Suppose that someone dared you to wear a button stating your annual salary, and you had to wear this button everywhere you went. Would you do it? Why or why not?

will look at societal attitudes about wealth and poverty and consider how cultural norms may change our definitions of those terms. We will see that labeling and stereotyping others in regard to socioeconomic status is as harmful as any other labeling and stereotyping, and as mistaken a notion.

Status and Stereotypes

Our socioeconomic status determines to a large degree the ways in which we are perceived by others. One can acquire prestige and popularity in a number of ways in our society. According to sociologist Tim North (1994), one important way to acquire prestige is by occupation. Employment as a laborer, for example, offers less prestige than employment as a surgeon. A stereotype we have about laborers is that they are less educated and have less wealth than, say, a physician. If we think before we make this assumption, however, we realize that a laborer may also *be* a surgeon, or a musician, or a writer. The surgeon may have more or less education, more or fewer skills, and be rich or poor. The great humanitarian Mother Teresa is a small, frail, elderly, unemployed woman who dresses very simply and lives in an impoverished section of the city, among the ill. By stereotypical standards, what should her socioeconomic status be?

Another way that we award status is by appearance, or dress (see Figure 7.2). We may make assumptions about a person's socioeconomic status based on whether he or she is wearing combat boots, cowboy boots, sneakers, or high heels. We may judge by the brand names on someone's clothing, or by what kind of hairstyle that person has. We all use these superficial clues to assess another's status, and they can be useful tools in learning about another person's style,

Figure 7.2
People often make assumptions about socioeconomic status based upon clothing.

taste, or job. Even that can be deceiving, however. A popular style among students a few years ago was wearing hospital greens. As we have seen in our examination of other kinds of stereotyping, it is always a mistake to judge by appearances. For instance, your grandmother might look at rock singer Axl Rose, of Guns 'N' Roses, and think that he is a poor, unemployed street person.

In the same way that prejudicial judgments may be made based on clothing or occupation, status is often assigned according to where people live (see Figure 7.3). One's address may be seen is an indicator of wealth or poverty. As evidenced by a popular television show of the early 1990s, *Beverly Hills 90210*, a zip code alone can be a kind of code for a way of life, or for a certain socioeconomic status which represents a lifestyle. Within that sequence of numbers is contained assumptions about how a group of people is likely to think, behave, dress, and interact. That zip code evokes in the viewer an assumption that the residents live in large houses, drive expensive cars, and have particular occupations. Similarly, an earlier television series, *Dallas*, focused on stereotypes about wealthy Texans, while *All in the Family* and *Roseanne* centered around lower class working families.

We can all enjoy these representations of socioeconomic stereotypes, and can recognize familiar patterns, but as with all other instances in which we judge individuals based on stereotypes, we risk making hurtful and harmful mistakes, and stand a good chance of impeding both the growth of our own understanding and the progress of society toward gaining the potential benefits of inclusion. Many university students might suffer from false assumptions based on zip code since they often live in lower rent areas, sometimes in crowded conditions and without the amenities they have at their parents' homes.

Figure 7.3
Where a person lives, and in what type of housing, is not always a reliable indicator of socioeconomic status.

Economic Issues Individuals May Face

- Whether or not to relocate in order to find work
- Paying taxes
- Whether to send a child to private or public school
- Whether or not to live in the suburbs or inner city
- How much to give to charity
- Which charities to support
- How to spend discretionary income
- Investment opportunities
- Planning for retirement
- Finding health care

Poverty

We can see the danger in making assumptions by looking at some statistics. According to the Bureau of the Census (1995), in 1990 more than 52 million Americans lived in areas where at least 20 percent of residents were poor. These areas are designated "poverty areas." However, most residents of poverty areas were not poor. Though these areas have high concentrations of poor people, statistics show that 69 percent were actually above the poverty line. More than half the people living in poverty areas were white.

In 1993, there were 39.3 million Americans below the official government poverty level. This number represents 15.1 percent of the nation's population. The poverty rate for children under 18 was 22.7 percent, and persons 18 through 64 years of age comprised 12.4 percent of the poor (Johnson, 1995). According to economist Peter Montague (1994), the United States has a higher incidence of poverty than other industrialized countries such as Norway, West Germany, and Japan, with exceptionally high infant and child poverty.

There is a disparity by race in terms of socioeconomic status. Most recent Census statistics show that non-Hispanic Whites had a poverty rate of over 12 percent, while Blacks had a rate of over 33 percent. The largest representation of people of other races was Asians and Pacific Islanders in 1993, and their poverty rate was over 15 percent (Johnson, 1995). It is important to note that though the poverty rate for Whites was lower than for those of other races and ethnicity, they still made up the majority of those suffering poverty in the latest census.

Families with a female head of household still made up a majority of people living in poverty. Families maintained by a woman alone had a poverty rate of over 35 percent as compared to married couple households with a rate of nearly 7 percent. If you were born in

First Person

I lost my job six months ago. My kids do not have any shoes, and I cannot afford to feed them every day. They don't complain, though. They're tough.

We live in an abandoned building. I found all of our furniture on the streets. We have a table, a chair, a sofa, and a bed.

I look for aluminum cans to sell for money. I worry about my kids when I am gone all day. I feel very sad all the time. I don't know where we will be tomorrow.

—Beth H., Atlanta

another country, you are over one and a half times as likely to be poor as a native-born citizen. Recent immigrants are over twice a likely to be in poverty (Johnson, 1995).

Poverty among those under age thirty has more than doubled during the past twenty years. Census surveys report that in 1990, eight American men aged 25-29 earned a total income of under $30,000 for every one making over that amount (Howe, 1993). There is a declining standard of living for this generation, but our attitudes about wealth have not adjusted.

Certain societal problems are directly attributal to poverty. Poverty breeds crime, disease, and illiteracy. Government attempts to address the problems of poverty in our society include such programs as Head Start (a pre-school program), the Food Stamp Program, the National School Lunch Program, the Low-Income Home Energy Assistance Program, Aid to Families with Dependent Children, and Supplemental Security Income (Bureau of the Census, 1995).

Come up with at least ten things that money can't buy.

Untrue Stereotypes of People Living in Poverty

- Are usually homeless
- Are prone to substance abuse
- Have below-average intelligence
- Are all unemployed
- Are lazy
- Are uneducated
- Choose to live in poverty
- Have no hope of escaping poverty
- Were born into poverty
- Have equal opportunity
- Are not law-abiding
- Are ineffective parents

Wealth

Our quality of life can be measured in various ways. The standard of well-being is different in a small village in Africa or Asia than it is in Atlanta. In thinking about our own socioeconomic status, we may consider the things we own, whether we can afford adequate housing, the safety of our neighborhoods, our ability to enjoy health and good nutrition, and our ability to earn a good income (Bureau of the Census, 1995). Clearly we are poor unless we have educational opportunities, health resources, and access to justice.

It is important to consider that the possession of money, though empowering in many ways, is not an indicator of an individual's personality, taste, lifestyle, or value as a human being. For instance, billionaire

Misconceptions of People Who Are Wealthy

- Are all hard workers
- Are all shallow
- Were born into wealth
- Are all employed
- Are all greedy
- Are never generous
- Are always law-abiding
- Are never victims of crimes
- Have little social consciousness
- Have great social consciousness
- Are effective parents
- Are well-educated

Howard Hughes became a recluse at a young age, by many accounts living much like a homeless person. Sam Walton, the super-wealthy founder of the WalMart stores, is said to have taken his lunch to work every day in a paper bag and in other ways continued to live as he had when he was poor. If we think of wealth as a quality of life, then it is possible to feel very wealthy without accumulating money. There is value in simple pleasures, in knowledge, and in service to others. Many of the world's richest and most powerful people have discovered this fact and valued life more than wealth.

Those who are considered wealthy in America often attend the best schools, have the finest medical care, and achieve professionally at the highest level due to status and connections. However, children and adults from the wealthiest group of citizens, like those who are poorest, are often mistrusted by the vast middle class of Americans. There is prejudice and resentment against both groups. Although wealth may bring with it the opportunity for a luxurious lifestyle, as it is often depicted on such television shows as *Lifestyles of the Rich and Famous,* it also brings problems, stereotypes, and stigmas. We must recognize that, no matter what their socioeconomic status, human beings all have the same physical, emotional, and intellectual needs.

Terminology

It is important to understand the following common terms describing people with socioeconomic differences:

Lower class. This refers to a segment of society which lives in poverty, sometimes called "the working poor."

Lower middle class. This refers to the largest segment of society, comprised mostly of white-collar workers employed in a variety of middle-income occupations.

Social class. This is an abstract concept which refers to a hierarchical layering of people into categories based upon their income, level of education, family name or bloodline, and influence.

Social status. This is determined by the prestige, esteem, and honor one is accorded within his or her own social milieu (Cushner, McClelland, & Safford, 1996).

Underclass. This refers to a segment of society which seems unable to take advantage of any mobility options and thus lies on the outskirts of the class system (Cushner, McClelland, & Safford, 1996).

Upper class. This refers to a small segment of society, the social elite, and is a position often inherited from one's family.

Upper middle class. This refers to a segment of society comprised mostly of highly educated professionals.

Working class. This refers to a segment of society comprised of blue-collar workers with low-paying industrial or service jobs.

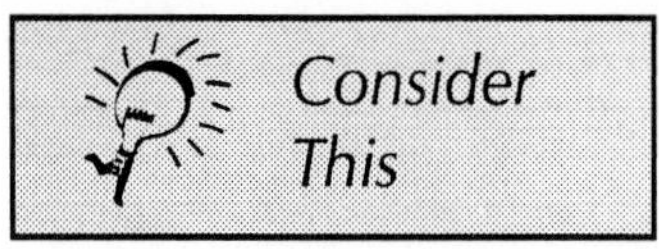

The United States Census Bureau has set the poverty threshold to a single dollar amount for all Americans. How might such a specific figure be misleading? For example, how could the same income level be comfortable for one family and grossly inadequate for another?

Support Services

Poverty is often related to a limited education. Local YMCA/YWCAs, Jewish community centers, adult alternative education centers, church organizations, and community clubs offer classes for little or no cost. Most states also have a community college system which includes a low-fee set of education and literacy courses. Most communities have a labor service agency that assists with job placement and referral to retraining programs. These agencies provide emergency assistance and case management for the homeless. These services can be found in the local Yellow Pages.

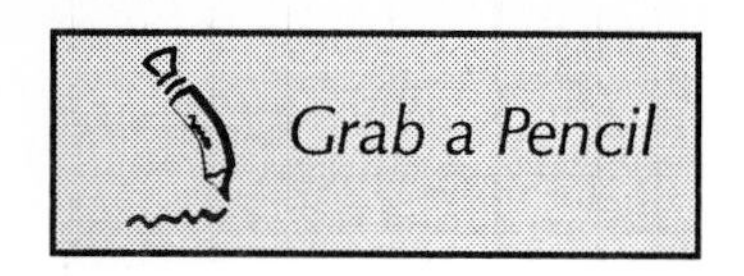

Brainstorm five disadvantages to being wealthy.

Interactions

Be careful not to stereotype people based on superficial aspects such as clothing, zip code, or occupation. Approach every individual respectfully.

Be charitable. Donate your excess to organizations which enhance life for those who have less.

Be active in your community. Give some of your time and energy to help guarantee equal access to justice and education for those who live in poverty.

Be instrumental in stopping poverty at its roots. Encourage children to learn to read, to be curious about life, and to participate in their own lives.

Be optimistic. Remember that poverty is reversible. Help one individual at a time. Generosity is not dictated by the size of your bank account. You can give your support, your knowledge, your sense of humor, and your love.

Become more aware of your own wealth. Appreciate the material advantages that you enjoy while increasing your understanding of what wealth really means. Poets, philosophers, spiritual and social leaders all teach us that money is not the measure of true wealth.

Get to know people who spread the wealth of human resources. Such organizations as Habitat for Humanity and Red Cross offer opportunities to help fight poverty and at the same time show that very valuable gifts can come from one's own effort rather than a checkbook.

Conclusion

President Lyndon Johnson declared a war on poverty in March, 1964, vowing to help replace despair with opportunity. Since then, statistics show communities are poorer and the gap between the wealthy and other Americans has widened. Opportunities for education, nutrition, and job-training are still absent in many areas of society. It's time to stop declaring a war and to start fighting it.

References

Bureau of the Census. (1995). Income and poverty. *Census home page.* [On-line]. Available: http://www.census.gov/ftp/pub/hhes/www/incpov.html

Cushner, K., McClelland, A., & Safford, P. (1996). *Human diversity in education: An integrative approach.* New York: McGraw-Hill.

Johnson, O. (Ed.). (1995). *1996 information please almanac.* Boston: Houghton Mifflin Company.

Montague, P. (1994). Economic trends. *Rachel's environment and health weekly.* [On-line]. Available: ftp.std.com/periodicals/rachel

North, T. (1994). Acquiring prestige and popularity. *The Internet and Usenet global computer networks: An investigation of their culture and its effects on new users.* [On-line]. Available: http://foo.curtain.edu.au/Thesis/Chap4b.html

Suggested Readings

Hodgson, B. F. (1911). *The secret garden.* New York: Harper Collins.
This book explains how friendship and love can bring happiness into the lives of people who have nothing.

Rose, S. J. (1992). *Social stratification in the United States.* New York: New Press.

This report analyzes data on wealth, race, martial and occupation, and income. It provides a vivid description of the U.S. social structure.

UNICEF. (1995). *The state of the world's children.* New York: Oxford University Press.

This report discusses the progress made since the 1990 World Summit for Children, providing ideas for where the world needs to be in order to fight the war on poverty.

Individual Activities

1. How would you categorize your economic status? Upon what criteria did you base your answer? Discuss below.

2. Forbes magazine publishes an annual list of the top 500 wealthiest people in the world. Who were the top three wealthiest people last year? How did they become wealthy? What do they do with their money? Discuss below.

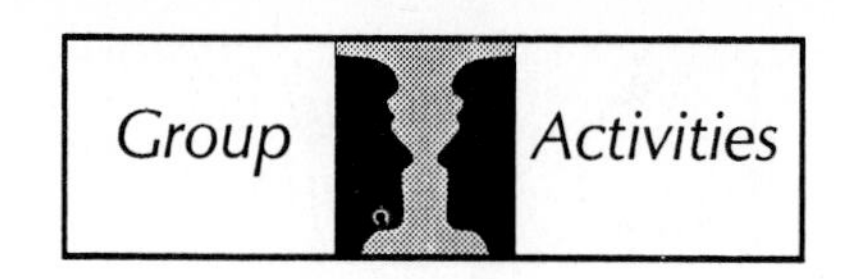

1. Each student should bring five pennies to class. Play a game in which each of you must give away as many pennies as possible to your classmates. As the game progresses, you will see that you cannot give away the pennies fast enough because more pennies keep coming back to you. Our economy is described in terms of "fluidity"—the constant flow of money among individuals. Being affluent doesn't just mean being rich in dollars. It means being rich in generosity as well.

Reactions:

2. As a group, list five characteristics of people who are wealthy and five characteristics of people living in poverty. Discuss why you thought of these common stereotypes.

Reactions:

Reflection Paper 7.1

Children are the fastest growing segment of the population living under poverty. Identify some of the contributing factors to this situation. How can we ever hope to break out of this tragic cycle?

Reflection Paper 7.2

“Money makes the world go ‘round.” What is your reaction to this quotation?

Notes

Chapter 8 Physical Differences

Chapter Sections

- Introduction
- Types of Physiognomy
- Size and Stature
- The Importance of Inclusion
- Terminology
- Support Services
- Interactions
- Conclusion
- Individual and Group Activities

"The human features and countenance, although composed of but some ten parts or little more, are so fashioned that among so many thousands of men there are no two in existence who cannot be distinguisged from one another" —Pliny the Elder, Roman naturalist

Objectives

- Distinguish between physical and orthopedic impairments.
- Explain ways society thoughtlessly discriminates on the basis of size and stature.
- Identify the various types of physical disabilities.
- Identify some causes of physical disabilities.

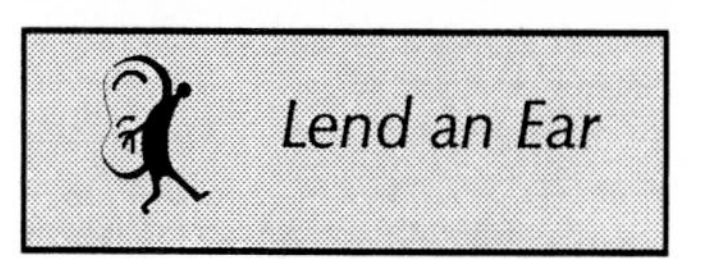
Lend an Ear

"One man's beauty is another's ugliness."
—Ralph Waldo Emerson, American poet

Introduction

Ideas about which physical attributes are desirable vary from culture to culture and from time to time. Our concepts of beauty are very complex and complicated. We are influenced by art and advertising, among other things, in our attitudes toward physical differences.

We know that all of us are different from each other and that people come in many sizes and shapes. Every day, babies are born who have physical disabilities. Many others of us suffer disabling accidents and illnesses. Still, society often poses challenges for people with physical differences because we tend to see the difference first and the person second. We live in a world which focuses on sameness. When physical differences are shown on television, in movies, or in advertisements, they are often the subject of humor, such as the "*fat* lady," the "*short* man," or the "*ugly* child." Comedian John Candy's weight, singer Barbara Streisand's nose, and actor Danny DeVito's stature are all well known topics of humor, along with countless others. Even actress Elizabeth Taylor, universally thought to be one of the most beautiful women alive, has been derided and ridiculed when she gained weight.

Figure 8.1

I am different.
This calls for a celebration!

Types of Physiognomy

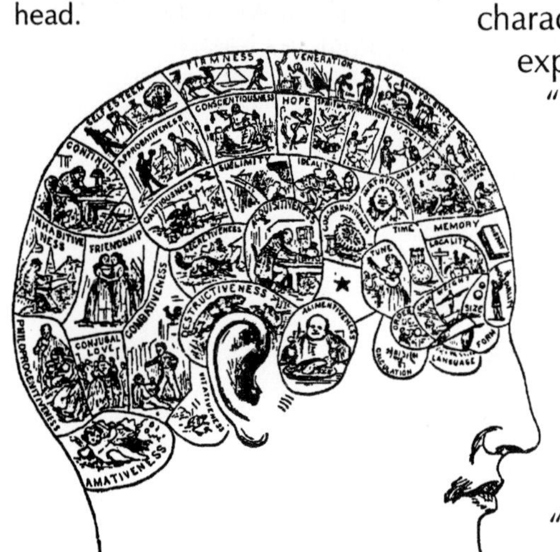

Figure 8.2 Phrenology sought to determine a person's character by analyzing bumps on the head.

The practice of reading a person's character from his or her facial and bodily form is not new. It dates back to the ancient Greeks and is still found in some parts of the world. We often judge a person's character by facial qualities, and our language expresses this. We say: "He has an open face," "She gave a frank look," "He had a sinister appearance," "She gave a furtive glance." The age-old art of interpreting physiognomy took all sorts of physical characteristics into account: one's stature, build, posture, forehead, nose, ears, chin, eyebrows, eyes, cheeks, mouth, and hair. Bumps on the head were also interpreted, in a pseudo-science called "Phrenology" (see Figure 8.2), as were lines on the palm of the hand, in "Palmistry." Today, we never make judgments on such superficial bases. Or do we?

People do sometimes fall into general "body types," and certain characteristics *are* associated with physical features. The study of this phenomenon is called morphology. For thousands of years and up to the present day, the traditional medical practitioners of China and India have relied upon morphology to classify and treat their patients. Do you fall into one of these three morphological categories?

Type 1	**Type 2**	**Type 3**
Moderate frame	Thin frame	Thick frame
Tend to be skinny	Moderate weight	Tend to be overweight
Dry hair	Soft, oily hair	Thick, wavy hair
Small eyes	Penetrating eyes	Big, attractive eyes
Very active	Moderately active	Lethargic
Irregular sleep	Little but sound sleep	Heavy sleep
Variable appetite	Good appetite	Slow but steady appetite
Restless mind	Sharp intellect	Calm, slow mental activity

Some people very clearly fall into one of these three categories. Others exhibit combinations of these traits. Looking at these charts, we can see that we cannot always take a person at "face value," nor should we be quick to judge others according to physical attributes. Surface differences don't necessarily reveal the important qualities of the person inside.

How comfortable are you with physical differences? Have you ever avoided greeting someone because of his or her physical dissimilarity from you? Was it because you thought that person might be unable to communicate, or perhaps because you were uncomfortable about the difference itself? Sometimes people feel embarrassed or even frightened when they encounter a physical difference. They might feel vulnerable, believing that a disability is contagious. They may behave self-consciously, not knowing how they should act, or how the person with disabilities will act. They may be worried that something strange might happen. They may also be curious, wondering what caused the disability, how the person functions, and what it would feel like to live with a disability (Berry, 1996).

Clearly, each of us is compelled to help others overcome ignorance and fear. Negative, destructive attitudes cause some people to exclude others from groups and learning experiences, thereby depriving everyone of enriching relationships. Thomas Bergman, a Swedish advocate for people with disabilities, points out that we, ourselves, set up social barriers which turn those with disabilities into handicapped people. When people who use wheelchairs must sit at the bottom of a step to a library or a theater, or those needing crutches or canes must walk where there is snow or ice, then they are handicapped (Bergman, 1989).

As a society, we have realized that people who have disabilities need the opportunity to develop to their fullest potential. As with the discrimination we have discussed in earlier chapters, our best hope for positive

Ask a *Friend*

Regarding people with physical disabilities, what do you consider our best hope for positive change in the years to come?

Issues that People with Physical Differences May Face

- People seeing the difference first and the person second
- Self-esteem and self-image
- Assaults on one's dignity
- Societal definition of beauty
- Labeling
- Public accommodations/ equal access
- Job discrimination
- Social, physical, or emotional isolation
- Finding support
- Ridicule from others
- Finding acceptance
- Relationships
- Harassment
- Alienation
- Overcoming stereotypes
- Ignorance and fear

Imagine wearing a sign on your forehead that reads, 'I was an abused child,' or 'I have problems with my boyfriend,' or even 'I am afraid of the dark.' Your classmates would stare at you and make snide remarks. They might laugh as you walk by. The luxury of those problems is that they can be hidden in your mind, and the note on your forehead doesn't really exist.

People who are overweight wear their problem on the outside, for all the world to see. Fat. It's a three letter word that I am pretty familiar with. I am not looking for sympathy, but if losing weight were easy I would have done it by now.

I want the world to realize that I know I am overweight. Don't censor conversation in the classroom when I walk in. Don't avoid sharing your bag of potato chips! Let me make my own choices, and when I am ready, if I am ready, then I will take on the task of losing weight.
—Tracy A., student

change lies in educating our children about physical differences. We must foster in ourselves and in our students the idea that "respect for what people can do [must] take the place of pity for what they cannot do" (Bergman, 1989, p. 7).

Size and Stature

Body size is obviously variable, but people whose size significantly varies from the average still may find themselves isolated physically and emotionally. Those who are obese, or are unusually thin, small, or tall may be subject to harassment or discrimination because of physical differences. It is important to educate yourself about some causes of these physical differences.

Approximately 38 million Americans are significantly heavier than average (National Association to Advance Fat Acceptance, 1995). People who are obese are discriminated against in education, employment, and access to medical care and to public accommodations. They are the victims of cruel and tasteless jokes and constant assaults on their dignity.

Most weight problems appear to be hereditary. Scientists have recently isolated a "fat gene" which could genetically predispose a person to be overweight. More and more doctors are now advocating a stable weight, sensible diet, and exercise for fitness rather than equating thinness with health. In any case, stereotyping people by body weight and attaching a stigma to size is clearly an act of ignorance as well as one of cruelty.

Thinness, like obesity, may be associated with an individual's metabolism rather than with an eating disorder. Some people may be genetically predisposed to thinness. Anorexia, a lack of appetite and/or inability to eat, is a specific disease. Labels such as "anorexic" should never be used by anyone except a trained diagnostician.

The pituitary gland releases growth hormones which stimulate cell reproduction and help to increase body growth (see Figure 8.3). Deficiencies or excesses of growth hormone in childhood can cause extreme variations in a person's height. Oversecretion of growth hormone, usually caused by a pituitary tumor, results in *gigantism*. An individual with gigantism may reach a height of seven to eight feet. Height which is below the fifth percentile on an incremental chart is called *dwarfism*. The causes of dwarfism can be environmental or genetic. Today, genetically engineered growth hormones can be periodically injected in children to spur their growth.

Figure 8.3
Body growth is regulated by hormones.

If you find any of these physical differences difficult to encounter, or have been affected by the stereotypes portrayed in movies and television, remember that thinness as an ideal is very recent in history. Extra weight was thought of as a sign of affluence and health in the last century, and women were thought of as voluptuous rather than "fat." Keep in mind that a basketball player whose height is more than 7 feet is considered a star, not an exception, and an Olympic gymnast who is four and one-half feet tall is not a subject of ridicule for his or her difference. Jockeys are prized for their small size and are paid very well for being "different." When you notice a physical difference, whether it is one of size or stature, remember to look at the person instead of the difference.

Have you ever felt left out? When you were younger, possibly even now, were you ever the last one picked to play a sport? Did people ever laugh at your abilities while you tried to do something such as dancing, reading, drawing, or music? How might a person with physical differences feel when unable to do a specific task well or when alienated from the group?

The Importance of Inclusion

Consider the ways in which our society thoughtlessly discriminates on the basis of size and stature. Desks and chairs often are designed only for "average" height and weight. Water fountains, toilets, sinks, and doorways in public buildings are also built for the needs of the "average" person. A policy of

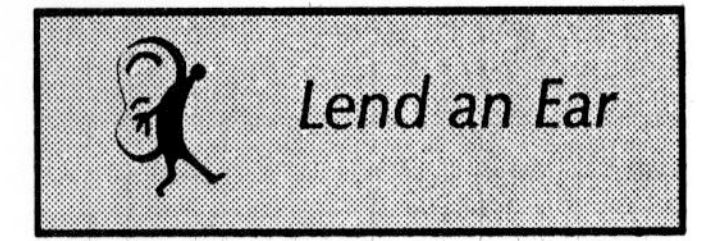

We never know how
high we are
Till we are called to rise
And then, if we are
true to plan
Our statures touch the skies
—Emily Dickinson, American poet

inclusion and understanding is of personal interest to each of us because at some point in our lives, most of us will have to cope with a physical disability. Physical disabilities are problems caused by injuries or conditions of the central nervous system which interfere with mobility, coordination, communication, or behavior. Orthopedic impairments may be caused by such congenital abnormalities as clubfoot or absence of a limb, or by impairments such as cerebral palsy, amputations, burns, and fractures. As we age, many of us suffer disabling conditions such as arthritis, and one in every twenty babies is *born* with some type of disability. Therefore, when we think of the necessity of providing access and support to "people with disabilities" in our schools, we are most likely thinking of ourselves and our families.

Terminology

Familiarize yourself with the following terminology relating to common physical differences and impairments.

Acquired disability. This is a condition resulting from illness or injury.

Arthritis. This is a painful disease which restricts movement of the joints.

Atrophy. This refers to a wasting away of muscles which are unused, such as after an injury.

Brain damage. This refers to a brain defect which prevents certain movement or affects thought, vision, or hearing.

Cerebral Palsy. This is a non progressive disorder of movement or posture that begins in childhood and is caused by a malfunctioning of or damage to the brain.

Congenital disability. This is a condition a person has from birth which limits ability.

Hemiplegia. This refers to paralysis of one side of the body.

Impairment. This is a loss of strength, feeling, or ability to move.

Monoplegia. This refers to paralysis of one extremity.

Multiple sclerosis. This is a degenerative, progressive disease of the central nervous system which involves hardening of the brain tissue. The symptoms include weak muscles, spasticity, balance difficulties, severe numbness of the appendages, and paraplegia. Symptoms may have periods of remission.

Muscular Dystrophy. This is an inherited condition resulting in progressive weakening and deteriorating of muscular tissue.

Orthopedic. This has to do with the bones, muscles, and joints used in movement.

Paraplegia. This refers to paralysis of the lower part of the body, usually caused by damage to the spinal cord.

Physical therapy. This method of treating a disability or injury may involve exercise, massage, medication, and education.

Prosthesis. This is an artificial replacement for a missing body part, such as an arm, hand, or leg.

Quadriplegia. This refers to paralysis of the body from the neck down.

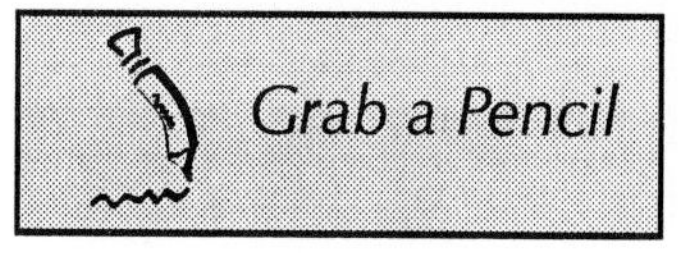

Check newspaper and magazine reviews and ads for restaurants and theatres to see if there is information about wheelchair accessibility. If there isn't, write to the managers of those establishments.

You are a world-famous inventor and one day you see a little child in a wheelchair. This sight affects you deeply and at this moment you decide to dedicate your life to inventing items that will be of use to people of all ages who are physically disabled. What are two or three inventions you will create and how they will help people who are physically disabled?

Spasticity. This refers to sudden muscle contractions that a person cannot control.

Spina Bifida. This is a congenital disorder which causes an opening of the spinal column in the lower back. It is the most common birth defect in North America and is the major cause of paraplegia in young children (Ward, 1988).

Support Services

Many people with limited locomotion or motor functions require supportive services. People with mobility disabilities use braces, canes, and wheelchairs for mobility assistance. Elevators, ramps and lifts, specially-designed buses, and canine companions also help people to live independently.

Examples of assistive technology include beeper baseballs, motorized wheelchairs, adapted skis, bathing and grooming devices, electronic communication devices, specialized hand tools, computer workstations, memory aids, spelling aids and educational software.

Interactions

Treat all people alike. Don't refer to those smaller in stature as "cute" or "shorty." Don't expect more mature behavior from a child of above average height for his or her age.

Extend your hand first when you are being introduced to anyone. Make eye contact. Failure to do so excludes those with disabilities or differences.

Respect wheelchairs as part of the user's space. Don't assume that he or she wishes to be moved. Always ask whether your assistance is needed.

Be an attentive listener to people who have a disability. Never pretend to understand. Ask them to repeat when necessary.

Put yourself in the line of sight of the person to whom you are speaking. Don't force someone in a wheelchair to strain in order to see you.

Never use labels to describe anyone. The disability is not the person and should never be used to describe him or her.

Do not hesitate to touch people with disabilities. Interact with people with disabilities as naturally as you would with any other people.

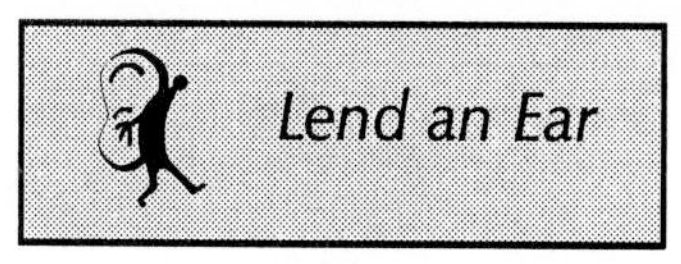

"Example is not the main thing in influencing others. It is the only thing." —Dr. Albert Schweitzer, Alsatian missionary

Conclusion

When you meet someone who uses a wheelchair, an individual with a prosthetic arm or leg, a person who is missing a limb, or an individual whose body is scarred due to a severe burn, challenge yourself to see beyond that person's difference. Whether different due to body size or disability, that person deserves respect. If your individual comfort levels seem to be affected by physical differences, take steps to interact more often with persons who look different. You will discover that the "difference" will fade away.

References

Bergman, T. (1989). *On our own terms: Children living with physical disabilities.* Milwaukee: Gareth Stevens Children's Books.

Berry, J. (1996). *Good answers to tough questions about physical disabilities.* Chicago: Children's Press.

National Association to Advance Fat Acceptance. (1995). *Fighting size discrimination and prejudice.* Sacramento, CA: Author.

Ward, B. R. (1988). *Overcoming disability.* London: Franklin Watts.

Suggested Readings

Beisser, A. (1989). *Flying without wings.* New York: Doubleday.
This is a first hand account of living with a physical disability.

Bergman, T. (1989). *On our own terms: Children living with physical disabilities.* Milwaukee: Gareth Stevens.
This is an excellent children's book presenting pictures of children with various disabilities in everyday situations.

Brimer, R. (1990). *Students with severe disabilities: Current perspectives and practices.* Mountain View, CA: Mayfield Publishing Company.
This book provides excellent global introductions to many severe physical disabilities. A historical perspective and theories related to causes and cures are included.

Brown, T. & Ortiz, F. (1982). *Someone special just like you.* New York: Henry Holt.
This book was written to give children and adults a better understanding of disabilities.

National Association to Advance Fat Acceptance. (1995). *Fighting Size Discrimination and Prejudice.* Sacramento, CA: Author.
Everything you need to know to dispell myths and stereotypes about being overweight is included in this publication.

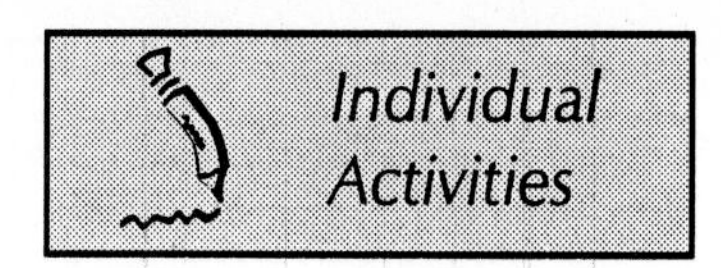

1. There are many people who have difficulty performing tasks that are usually taken for granted. This activity will give you a perspective on how these people feel about and deal with these tasks. Hold one arm behind your back and try to do the following activities: tie your shoes, open a can or a bottle, get dressed, eat something, and play a sport that normally requires two hands. Now, remember, only one hand and no cheating.

Describe how well you did. Also describe the difficulties you encountered.

2. Assume that you are confined to a wheelchair and are about to start your first day in college. But wait. Your college campus is not set up for wheelchairs. How do you feel?

List some things that may be obstacles to you.

Group Activities

1. As a class, take a trip to the local mall or to your college's student union. Today, you will have a physical disability. You are to interlock your fingers from both hands and place them on top of your head. If you feel silly, remember that everyone else is doing it also! The task assigned to you is to pair off in groups of two with only one person at a time acting as the person with the disability. While disabled, have your partner feed you and wash your face. Also, play a video game, bowl, or play some other type of game. As you walk around the mall or union, open doors for yourself and be as independent as possible.

Reactions:

2. A town meeting is being called to order and your class is the community. Five people need to be appointed to the town council to weigh the information presented to them by the community—namely, the rest of the class. The issue of concern in your town is a petition being circulated to keep overweight students out of the local public schools. Divide the class into sides: one to argue for these students to be given the right to attend the school, and the other side to argue against the action. The town council will be the decision-making body. Before the meeting begins, take 5 to 10 minutes and brainstorm ideas supporting your side of the issue. Allow time for opening comments, the actual arguments, rebuttals, and closing comments. After the town council members have made a decision on how they will vote, ask for their reasons.

Reactions:

Reflection Paper 8.1

Reflection Paper 8.1

Assume that you just met a co-worker who has a physical disability and uses a wheelchair. Will you be comfortable? Why or why not?

Reflection Paper 8.2

Our society's standards of physical beauty change over time. In the seventeenth century, for example, obesity was the ideal. How are today's ideals beneficial (or not) to our society?

Notes

Chapter 9

Learning Differences

Chapter Sections

- Introduction
- The Culture of Learning
- Identifying the Problem
- Categories of Learning Disability
- Causes of Learning Disabilities
- Information Processing
- Terminology
- Support Services
- Interactions
- Conclusion
- Individual and Group Activities

"Learning is not attained by chance, it must be sought for with ardor and attended to with diligence" —Abigail Adams, First Lady

Objectives

- Explain the primary way learning disabilities are identified.
- Distinguish one characteristic shared by all people with learning disabilities.
- Specify four possible causes of learning disabilities.
- Identify five specific types of learning disabilities.

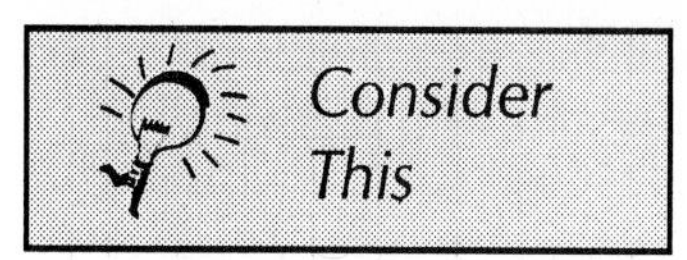

We all take advantage of learning aids to make the job of processing information easier. What are some learning aids that you frequently use, and why?

Introduction

When you learn a new game, what do you do first—read all the rules, watch others play, or just start playing and learn as you go? When you are introduced to new information, do you remember it better if you see it on paper, hear it, write it, or say it aloud? No two people learn in exactly the same way. People with learning differences may be good at processing information in one way but not in another.

We frequently rely upon technological aids to make learning easier—calculators for difficult math calculations, word processors to check our grammar and spelling, and tape recorders to record or play back information (see Figure 9.1). Books on tape often sell as well as their printed counterparts. Computer programs help us to master foreign languages, learn to draw, and balance our checkbooks. Clearly, many of us take advantage of ways to augment our learning. And many of us are underachievers, whether academically or professionally. People with learning disabilities, though, may not have poor achievement because they have learned good coping strategies. Learning disabilities can be related to disorders of thinking, learning, and sometimes of communication. They are called by different names, and there are many

Figure 9.1
Everyone augments their learning with technology and other aids.

Untrue Stereotypes of Persons with Learning Differences

- They can't understand
- They represent a small minority
- They are not trying hard enough
- They all have brain damage
- They are "slow"
- They always have trouble academically
- They cannot read or write
- Their social and emotional well-being is not affected
- They usually outgrow their learning disabilities
- They are stupid
- They need special classes
- Teachers can always identify children with learning disabilities

types, but we will use the general term *learning disability* to describe the exceptionality in general, and briefly discuss a few of the specific types in more detail.

A learning disability is an invisible disability. Because people with learning disabilities are in the normal range of intelligence and may function perfectly well in most contexts, it is sometimes difficult to assess their problem. People with learning disabilities look and act like everyone else for the most part. Until they are asked to read, write, solve mathematical problems or demonstrate some other academic skill, or until they exhibit the inability to organize learning tasks, it may not be apparent that there is any problem at all. A person with a learning difference may be able to recite sports statistics, remember song lyrics, or show artistic talent, for example, but in some particular areas be unable to perform adequately.

It may be frustrating for both the person with a learning disability and for that person's family, friends, and teachers to understand what is happening and to cope with the problem, especially until such time as the difficulty is properly diagnosed. The reason for the frustration is that frequently the person with a learning disability has average to high intelligence and cognitive ability. This chapter seeks to introduce the many facets of learning disabilities so that you will be better equipped to understand friends and family members who learn differently.

The Culture of Learning

People with learning differences are frequently successful because they have developed techniques for camouflaging their differences and getting by in the world. Many of your colleagues and peers likely have a learning difference, though you may be unaware of it. For example, a business executive who has trouble reading or writing might dictate letters and have a secretary type them. In this case, the potential disability is circumvented. Another person might rely upon digital watches to tell the time, or calculators to solve math problems. Such a person can accurately tell time and solve math problems by using assistive technology. A traveler lost on the road might ask for directions rather than consult a map, again taking advantage of alternate sources of information. Someone may even pretend to read a book or newspaper while sitting on the city bus in an effort to fit in with the crowd. Indeed, an entire culture of learning has arisen to aid people with learning differences in succeeding in life and going unnoticed.

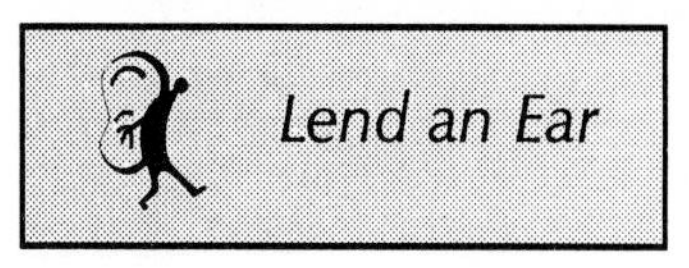

"Learning is but an adjunct to ourself." —William Shakespeare, playwright

Identifying the Problem

Learning disabilities are the newest categories of exceptionality, having been identified only since the 1960s by educators and parent groups. Estimates of the number of students with learning disabilities range from 15 to 50 percent. Many students are not diagnosed with learning disabilities until college level. Have you ever had trouble following a lecture, understanding a reading assignment, or keeping up with the pace of a classroom activity? Three to five percent of college students are diagnosed with learning disabilities, even though they experienced no difficulties in their earlier school experiences. It has been estimated that over one million children will have received stimulant medication in the 1990s in attempts to treat learning disorders (Associated Press News Service, 1989). Of all categories of exceptionality, learning disability (LD) affects the greatest number of individuals and requires the greatest number of instructional personnel. It also is the subject of the most disagreement about definition and diagnosis, as well as treatment methods and educational services.

One major problem with people who live with learning disabilities is not their skills but their self-concept. It is difficult to advance when you are constantly thinking that you are not as good as everyone else. If you had a child with learning disabilities, what would you say or do to get the lower expectation out of your child's mind?

Some exceptionalities, such as race, gender, physical disability, size, or age, for example, are obvious because of their characteristic features or behaviors. Learning disabilities are very difficult to identify because individuals with LD are more like those without it than not. There is an extremely broad range of problems associated with learning disability, and its effect varies in degree from very mild to severe. The primary way that the exceptionality of LD is identified is by comparing a person to other people in his or her age bracket to determine if there is a discrepancy between expected achievement and actual performance. This discrepancy model is the basic method for identifying a person with a learning disability. For example, if an eight-year-old girl is not performing at the average or expected level for her age, IQ, educational opportunity, and ability, her poor performance is most likely due to some sort of learning disability. The girl is also likely to be exhibiting some social or behavioral problems as a result of the way the LD affects her feelings and/or the responses of the people around her.

Categories of Learning Disability

Let's examine some generalities about learning disorders and disabilities. Traditionally, perception, memory, and attention have been areas of concern, and educators, doctors, and psychologists have all had their different ways of assessing and addressing the problems. For example, educators may use such terms as *specific learning disabilities*. Psychologists may talk about *perceptual disorders* and *hyperkinetic behavior*. Speech and language specialists use such terms as *aphasia* and *dyslexia*. Doctors may use such labels as *dysfunction, impairment,* or *brain injury*.

We can say as lay persons that individuals with LD have, at some point in their lives, all experienced poor achievement in some context, and that that poor achievement was associated with learning needs. Since performance and achievement affect self-concept, and learning disabilities may inhibit the development of social and interpersonal skills as well, often individuals with LD may suffer emotionally and socially, though this is not always the case.

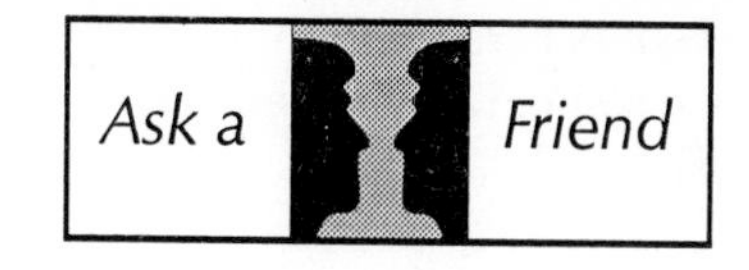

How might you encourage a positive attitude toward the exceptionality of LD among your friends, family, and neighbors?

Causes of Learning Disabilities

The causes of learning disabilities are the subject of much examination and debate. Prenatal research shows that a mother's use of tobacco during pregnancy may have damaging effects on the unborn child (National Institutes of Health, 1993). Mothers who smoke during pregnancy are more likely to bear smaller babies. This is a concern because small newborns, especially those weighing less than 5 pounds, tend to be at risk for a variety of problems, including learning disorders.

Alcohol also may be dangerous to the baby's developing brain. It appears that alcohol may distort the developing neurons. Heavy alcohol consumption during pregnancy has been linked to Fetal Alcohol Syndrome, a condition that can lead to low birth weight, intellectual impairment, hyperactivity, and certain physical defects. Any alcohol use during pregnancy, however, may influence the child's development and lead to problems with learning, attention, memory, or problem solving.

Drugs such as cocaine—especially in its smokable form, "crack"—seem to affect the normal development of brain receptors. These specialized neural cells help to transmit incoming signals from our skin, eyes, and ears, and help to regulate our physical response to the environment. Because children with certain learning disabilities have trouble understanding speech sounds or letters, some researchers believe that LD may be related to faulty receptors (see Figure 9.2). Current research points to drug abuse as a possible cause of receptor

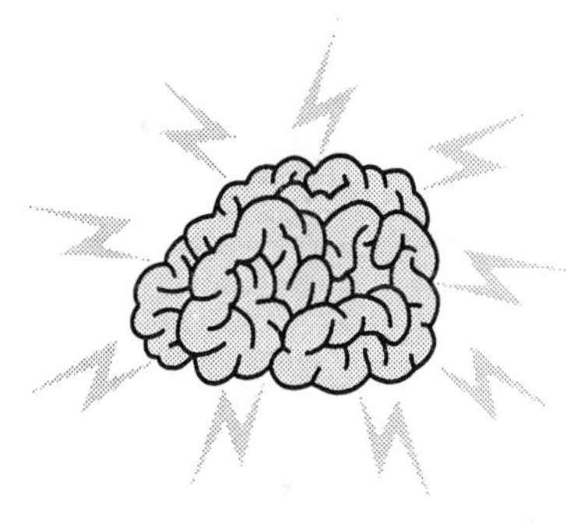

Figure 9.2
Drugs and alcohol impair the normal development of brain receptors.

Issues Persons with Learning Differences May Face

- Receiving appropriate service
- Ridicule from others
- Identifying the learning difference
- Classroom problems
- Mainstreaming v. non-mainstreaming
- Finding support or acceptance
- Finding role models and mentors
- Relationships
- Labeling
- Finding the best treatment
- Instructor's perceptions
- Adaptable or appropriate equipment
- Acknowledgment of difference
- Adaptation of classroom teaching strategies to adjust for learning differences
- Getting a job where the boss accepts or is willing to recognize issues related to learning challenges
- Others considering them slow or stupid
- Feelings regarding their need for assistance or different educational strategies

damage (National Institutes of Health, 1993).

Genetic abnormalities (passed on from parent to child) are another possible cause of learning disabilities. Complications during delivery, such as when the umbilical cord becomes twisted and temporarily cuts off oxygen to the baby, can also impair brain functions and lead to LD.

For a year or so after a child is born, fragile new brain cells and neural networks continue to be produced. Neuroscientists are identifying certain environmental toxins that may lead to learning disabilities, possibly by disrupting early childhood brain development or brain processes. Cadmium and lead, both prevalent in the environment, are becoming a leading focus of neurological research. Cadmium, used in making some steel products, can be carried from the soil into the foods we eat. Lead was once common in paint and gasoline—it is still present in some water pipes. A study of animals, sponsored by the National Institutes of Health, revealed a connection between exposure to lead and learning difficulties. In the study, rats exposed to lead experienced changes in their brainwaves, slowing their ability to learn. The learning problems lasted for weeks, long after the rats were no longer exposed to lead (National Institutes of Health, 1993).

Other environmental factors, including lack of reinforcement for learning by parents, radiation stress, fluorescent lighting, poor nutrition, food additives, and unshielded television tubes have been targeted as possible causes of LD. Clearly, learning disabilities have different causes, and any one disability may have many causes. Scientific research into this complex exceptionality is on-going, though no one factor has been proven conclusively.

Information Processing

Clearly, not all learning problems are actual disabilities. There is a continuum of learning differences, ranging from general learning style preferences to specific neurological damage (YMCA, 1996):

Learning Style Preferences
↓
Learning Differences
↓
Learning Problems
↓
Learning Disabilities
↓
Brain Damage

You may not always be able to diagnose exactly where a person falls on this learning continuum. However, by noticing the strategies that other people use to process information and to solve problems, you can discover a great deal about how they think and learn (YMCA, 1996). There are three basic categories of information processing: 1) visual, 2) auditory, and 3) tactile/kinesthetic. Though each person combines these modes to some extent, one mode will usually dominate as the primary learning style. Consider the following general characteristics (derived from YMCA, 1996) as you observe your own and other people's learning styles:

Visual Learner

- Scans everything; wants to see things; enjoys visual stimuli (see Figure 9.3)
- Stores visual images
- Enjoys shapes, colors, patterns, maps, pictures, diagrams
- Can recall words after seeing them a few times
- Has difficulty in lectures

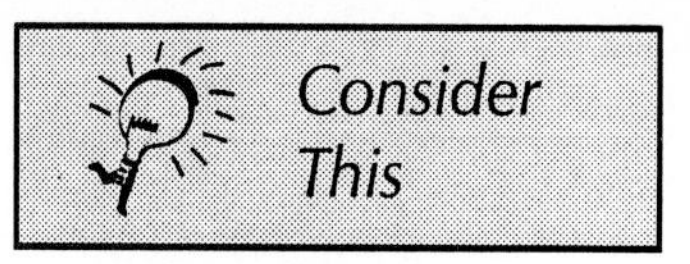

When you experience a problem, or see a problem that someone with LD is having, focus on the environment rather than on the learner. For instance, try to express directions more clearly, or in another way, or try demonstrating instead of telling the instructions.

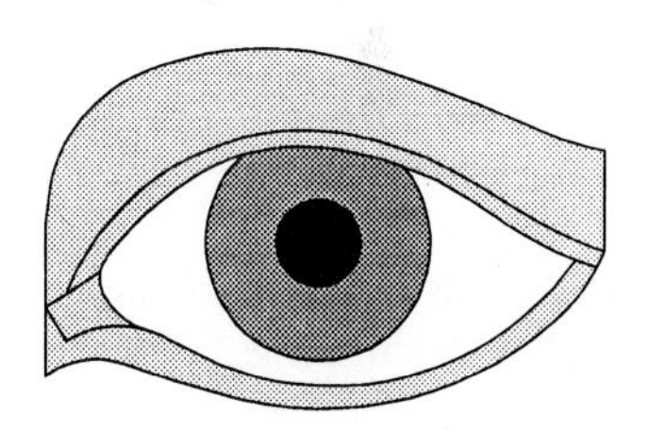

Figure 9.3
Visual learners want to see images.

- Daydreams
- Can vividly describe visual details of a scene
- Often a good speller; internalizes spelling patterns quickly
- Prefers written directions/notes
- Often finds locations by landmarks

Figure 9.4
Auditory learners prefer oral directions.

Auditory Learner

- Good listener; often the "talker"
- Prefers oral directions (see Figure 9.4)
- Prefers lectures to reading assignments
- Likes to hear stories, poems, music
- Seldom writes things down
- Often repeats what has been said
- Often moves lips while reading
- Likes to study with music, the radio, or the television on
- Often has a good ear for music

Tactile/Kinesthetic Learner

- The "do-er"
- Needs to touch, handle, manipulate materials and objects (see Figure 9.5)
- Counts on fingers
- Good at drawing designs
- Often doodles while listening, especially on the telephone
- Needs to get up and move around in order to process information
- Good at sports
- Good at mechanics, using appliances, or tools
- Often enjoys hiking, jogging, other non-stationary activities

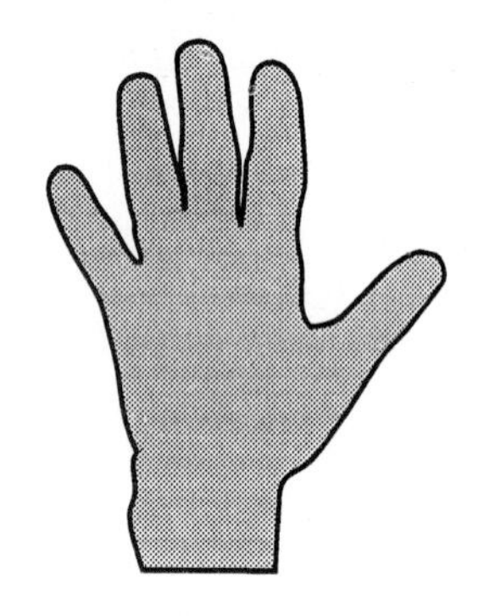

Figure 9.5
Tactile/Kinesthetic learners need to touch things.

An awareness of these basic learning styles will help you to adjust your expectations for people who learn differently than you do. Always remember that everyone is learning all the time—just in different ways.

Share with your friends and family the following bits of advice, which are appropriate for all types of learners (derived from Poor Richard, 1996):

- Consider everything an experiment.
- Learn by trial and error, and don't avoid the errors.
- Learning doesn't happen in class; it happens when you get home and look at the wall. Don't forget to set aside time for looking at walls.
- Be self-disciplined.
- Be a self-advocate.
- Learn from your mistakes. There is no win and no fail, there's only honest effort.
- Get good at something other than school-related work (like skateboarding or cooking).
- Don't try to create and analyze at the same time. They're different processes.
- Don't spend energy worrying; just get started and potential difficulties will work out.
- Subscribe to lots of magazines, the more pictures the better. Read what interests you, even if it's just one article.
- It's the process, not the product, that counts, because you can use it again and again and it transfers.
- Avoid doing school-related work under pressure. Always allow lots of extra time for things that are difficult, and for everything else, too.
- Learning opportunities are everywhere. The more you do, the more you learn. (See Figure 9.6).
- Read anything you can get your hands on. Comic books involve decoding just as great literature does. Read billboards and road signs.
- Write lots of letters.
- Look at movies carefully, often.
- Learn to ask questions without feeling stupid.
- Travel whenever and wherever you can.
- Give others some slack; it makes life easier.
- Give yourself some slack; it makes life easier.

Figure 9.6
Learning happens everywhere, not just in school settings.

First Person

When I was 67 years old, I had to learn to read. My husband had a heart attack and all of a sudden I was faced with bills from doctors and letters from family which I couldn't read, let alone answer. I was brought up in south Louisiana and my family spoke Cajun French. I never learned to read at all, but it didn't matter because my husband took care of all that. When he got sick it was terrible, though. I couldn't drive, and even after I took a cab to the hospital, I had trouble finding my way around. I couldn't read the signs, and at night there was almost no one to ask for directions.

I finally went to an adult education center where they assigned me a reading tutor. We studied one-on-one, and by the end of the first book I could read most traffic signs and could write my name and numbers so I could write checks. Now, after two years, I can read pretty much everything.

I like the feeling of being independent. I can find new recipes, read labels on cans, and look for sales at the store. Best of all, I don't have to hide anymore. I always pretended or avoided people in the past. Now I'm just like everybody else. It feels good.
—Marian J., Port Allen, LA

Terminology

Following are some broad, thumbnail definitions of a few categories of the exceptionality called learning disability. It is a good rule to be cautious about using these terms without the guidance of a professional. This list is offered solely for the purpose of introducing you to a few of the most commonly used terms.

Aphasia. Aphasia is the loss or impairment of the ability to use spoken or written language. An individual experiencing aphasia may not be able to recall the name of a familiar person or object.

Brain injury. An individual with brain injury is described as having an organic impairment. Such an impairment results in perceptual problems, thinking disorders, and emotional instability (Hardman, Drew, Egan & Wolf, 1993).

Dyscalculia. Impairment in arithmetic or computation skills is called dyscalculia. Individuals with dyscalculia may have trouble counting, writing numbers, or solving even simple addition or subtraction problems because they are unable to relate to the numerical figures.

Dysgraphia. Impairment in the ability to write is called dysgraphia. Individuals with dysgraphia may know how to spell a word but be unable to write it on paper. They may also have difficulty with handwriting, word spacing, or letter formation, or they may write very slowly.

Dyslexia. Impairment in the ability to read is called dyslexia. Individuals with dyslexia may have trouble spelling certain words and may invert letters of the alphabet. They may also have difficulty with comprehension as a side-effect of the dyslexia.

Hyperactivity. This is a behavioral characteristic which is also often termed *hyperkinetic*. Hyperactivity is typically a general excess of activity. Individuals with hyperactivity may be described as fidgety, nervous, or distractible.

Minimal brain dysfunction. This classification has become outdated, but you may still come across it occasionally. A person who shows behavioral but not neurological signs of brain damage is sometimes diagnosed with "minimal brain dysfunction" or "minimal brain injury." Individuals with minimal brain dysfunction are often average or above average in intelligence.

Perception disorders. Perception problems for individuals with LD may include a wide range of difficulties, including visual, auditory, and haptic (touch, body movement) sensory systems (see Figure 9.7). Visual and auditory difficulties in LD should not be confused with the characteristics associated with hearing and vision exceptionalities. These perceptual problems are entirely different. They have to do with perception, rather than with hearing and vision, per se. An example would be clearly seeing a word on paper yet reversing the order of letters.

Figure 9.7
Perception problems may include disorders in the haptic, auditory, and visual systems.

Specific learning disability. A legal definition which has been incorporated into federal law is that specific learning disability is a "disorder in one or more of the basic psychological processes involved in understanding or in using language, spoken or written, which may manifest itself in an imperfect ability to listen, think, speak, read, write, spell, or to do mathematical calculations" (Individuals with Disabilities Education Act, 1990).

Support Services

Figure 9.8
Private tutors can help people discover how they learn best.

Private tutors are often the best solution for people with learning disabilities. Through observation, discussion, experimentation, and modeling, a tutor can assess how a person learns best and teach him or her how to study accordingly (see Figure 9.8). Adult education is provided by local schools and community colleges to teach basic skills, General Equivalency Diploma (GED) preparation, and vocational education. Volunteer tutors participate in adult literacy programs in which individuals with low reading and writing skills are placed with tutors for one-to-one private instruction. Churches and synagogues frequently provide tutorial services, and colleges and universities offer community education classes.

Interactions

Establish eye contact before giving or repeating instructions. Establishing eye contact helps to make a stronger connection between you and the person with whom you are communicating.

It is important to be patient and to remember that the difficulties of individuals with LD do not come from lack of motivation or intelligence. Think of the problem as a challenge to you to communicate more clearly.

Encourage persons with LD to seek assistance from agencies which offer special help. Newspapers, talking books, and many other services are available in most communities.

Act to secure treatment and protective legislation for those with LD. Offer your help and support.

Volunteer as a tutor for children or adults with learning disabilities. Many organizations provide training for volunteers at no cost.

Live by these rules of thumb (adapted from Poor Richard, 1996):

- Pull as much as you can out of those around you.
- Extend yourself.
- Be creative.
- Don't assess someone's capabilities based on IQ scores.
- Don't be afraid to make mistakes in front of others. If learning takes place through modeling, you must model the process of working things out, from scratch, mistakes and all.
- Assume that others are always doing their best.
- Remember that we all learn differently.

Encourage a positive attitude toward the exceptionality of LD among your friends, family, neighbors, and fellow students. They, like you, will certainly encounter and interact with many people in their lives who have learning disabilities. The information you share may help them to identify their own LD in some cases, as well.

Conclusion

This chapter has introduced an exceptionality which is complex and comprehensive, and so offers you special challenges. You have seen that the term *learning disabilities* is broad and generic and that it involves different specific problems. The study of this exceptionality is relatively new, and though a variety of disciplines such as medicine, education, and psychology have undertaken it, even the conceptual development and terminology related to LD are still in the formative stages. The field is growing at a rapid pace, and our understanding is increasing. At the present time we can say only that those individuals exhibiting the characteristics of learning disabilities are so varied that we cannot describe them with any one concept or term.

Meanwhile, educators have developed excellent interventions for students with learning disabilities. "Learning Strategies," developed by researchers at the University of Kansas, have become recognized for their ability to assist persons with LD to *learn how to learn.* Special educators, who are trained in Learning Strategies, have seen significant improvements among students with learning disabilities as the strategies have provided structured ways for these students to learn effectively.

One major problem with people who live with learning disabilities is not their skills but their self-concept. It's difficult to advance when you are constantly thinking that you are not as good as everyone else. Your ultimate responsibility is to dispel lowered expectations from people's minds.

References

Associated Press News Service. (1989, June 3). 750,000 children take stimulants researchers say.

Hardman, M., Drew, C., Egan, M., & Wolf, B. (1993). *Human exceptionality: Society, school, and family* (4th ed.). Boston: Allyn and Bacon.

Individuals with Disabilities Education Act. (1990). Washington, DC: U.S. Government Printing Office.

National Institutes of Health. (1993). Learning disabilities: Decade of the brain. [On-line.] Available: gopher://zippy.nimh.nih.gov:70/00/documents/nimh/other/learn

Poor Richard's Publishing. (1996). Tips and advice for students, teachers, and parents. In *Poor Richard's publishing home page*. [On-line.] Available: http://www.tiac.net/users/poorrich/tipsadvice.html

YMCA. (1996). Tutor tips: Learning differences and disabilities. In *Adult literacy homepage*. [On-line.] Available: http://libertynet.org/~ymcalit/lrndis.html

Suggested Readings

Hall, D. E. (1993). *Living with learning disabilities: A guide for students.* Minneapolis: Lerner Publications.

This guide describes various LDs such as attention deficit disorder, fine motor problems, and difficulties with visual information, and offers positive advice on how to cope.

Hampshire, S. (1982). *Susan's story: An autobiographical account of my struggle with dyslexia.* New York: St. Martin's Press.

This book presents one woman's first-hand experiences with dyslexia.

Ingersoll, B. D. (1993). *Attention deficit disorder and learning disabilities: Realities, myths, and controversial treatments.* New York: Doubleday.

This book examines the symptoms, development, prognosis, causes, diagnosis, and treatment of learning disabilities and ADD.

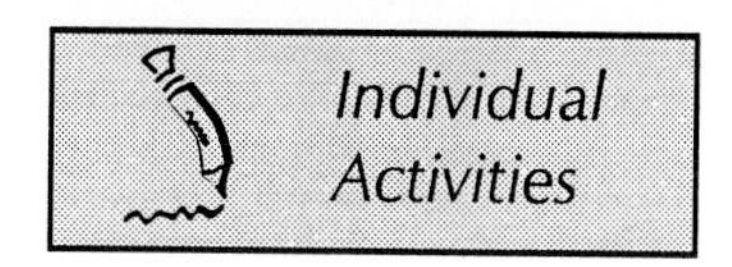

Read the following paragraph quickly:

> Down syndrome si a disaes dezirertcarahc by wol I.W., shrot and broad hands, and woleb average thgieh. Ti is more birth DNA there is no crue.

This is how an individual with a severe learning disability may see words. Did it take you longer than usual to read and understand it? Explain below.

Imagine having to deal with all reading materials in this fashion.

Now quickly write your name and address, but write all of the words backwards. Do you think people with learning disabilities feel the way you do now when faced with a difficult task? Write your reactions below.

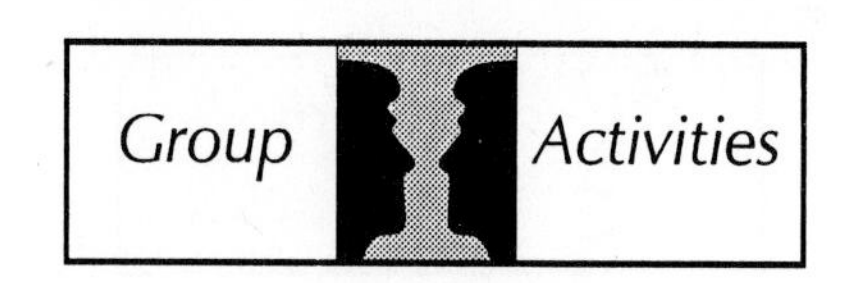

A learning disability may not just be trouble recognizing words or memory. It may affect one's ability to take notes. Have one member of your group read aloud a short paragraph from your textbook as quickly as a teacher would speak during a lecture. Take notes on this information in the space below. However, don't write with your preferred hand. Use the other! How well did you do compared to the other members of your group? Are you learning disabled?

Reactions:

Reflection Paper 9.1

Learning disabilities have been called "invisible disabilities." Explain how the invisibility creates special difficulties and frustrations for both the individual with the disability and those around him or her.

Reflection Paper 9.2

Reflection Paper 9.2

Discuss five concrete things you can do to assist an individual with a learning disability.

Notes

Chapter 10 Intellectual Diversity

Chapter Sections

"Almost all the joyful things in life are outside the measure of IQ tests."
—Madeleine L'Engle, author

Objectives

- Explain how intelligence is defined.
- Describe how intelligence is measured.
- Explain why intelligence assessments are controversial.
- Identify ten qualities associated with advanced intellectual ability.
- Identify the three criteria used to identify developmental delays.

Everyone has a unique talent which no one else on earth possesses. Discover and use that talent.

Introduction

Of all the differences we can observe between ourselves and others, perhaps the most important and the most interesting is intellectual difference. We use the terms *intellect* or *intelligence* to describe cognitive power—that is, the ability to learn and to process information (see Figure 10.1). The campus is an obvious center of learning and information processing, where students' intellectual abilities are constantly evaluated and challenged in oral discussions and written examinations.

We attempt to measure intelligence in many ways, including tests of creative thinking, verbal aptitude, academic knowledge, and psycho-motor skills. There continues to be disagreement among experts as to the validity and use of these tests, but they are in general use for assessment of special needs and for placement in schools and programs.

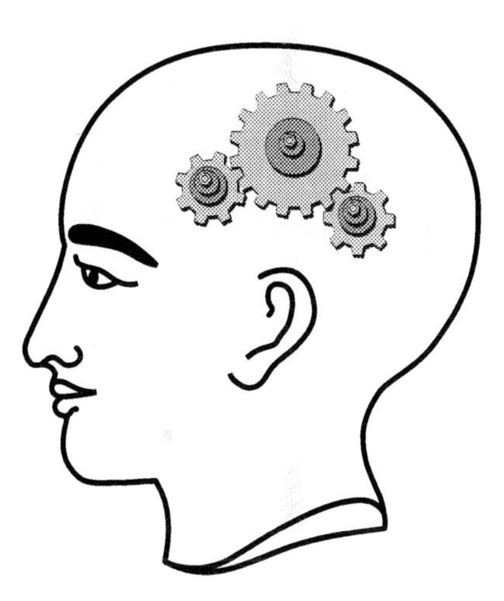

Figure 10.1
Intellect refers to one's ability to process information.

It is a mistake to categorize people solely on the basis of an intelligence assessment. Human beings have complex and surprising brains, and each of us has a broad range of aptitudes and potentialities. A test score alone can neither predict nor account for an individual's hidden potentials and abilities. At the extreme ends of

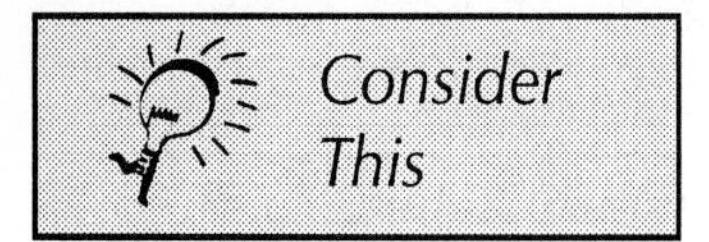

Consider a time when you were tested, whether it was an intelligence or aptitude test, or a school exam. Did you feel that the test score or grade was a true indicator of your knowledge? Why or why not? Did you feel that the score was a true indicator of your *potential*? Why or why not?

assessment scales, however, are the exceptionalities of giftedness and mental retardation. Giftedness describes advanced abilities, while being developmentally delayed (often called *mental retardation*) describes limited intellectual abilities.

Intelligence Testing

In order to establish norms for intelligence, researchers test random samples of populations which represent different ages, cultures, ethnicities, and genders. These studies are used to set a standard score, or norm. Individuals can then be compared to the norming group. You have no doubt taken any number of such tests during your school career: the G.R.E., S.A.T., A.C.T., and so on.

Society's definition of intelligence is not static. Through time, we have seen that many individuals realize potential achievement which is never indicated by any assessment scale. We also observe that many of us fail to perform at levels indicated as our potential. As we are able to see the truth that every human being has innate intelligence and therefore has value, we begin more and more to look for ways to enhance each person's potential.

Intelligence testing requires specialized training and is usually done by psychologists or others who are certified to administer the tests. Results of intelligence tests can be affected by cultural differences and other variables. Therefore, these results should always be understood in a limited context. They are designed to indicate areas in which a person might most benefit from specific instruction, for example. The Wechsler Intelligence Scale for Children and the Stanford-Binet Intelligence Test are typically given to evaluate intellectual and developmental levels.

Giftedness

One of the best arguments for studying human diversity is the definition we give to giftedness. Below is a list of qualities associated with advanced intellect. Which ones apply to you?

- Curiosity
- Flexibility
- Leadership ability
- Artistic talents
- Unusual empathy and concern for people
- Excellent verbal skills
- Genuine interest in learning
- Ability to grasp ideas
- Originality in thought
- Interest in designing, developing, and creating
- Ability to appreciate beauty
- Large vocabulary
- Persistence
- Ability to concentrate for long periods
- Read often
- Good memory
- Good judgment and logic

You can see that many of these qualities which we associate with giftedness indicate openness, multiple interests, and, in fact, diversity. Those who contribute most to our society are the ones who are themselves a celebration of diversity. Your campus is like a petri dish, swarming with living diversity, electrified with different impulses of intelligence (see Figure 10.2).

Educators often define giftedness in terms of a high intelligence test score—typically two standard deviations above the mean or norm—on individual or group measures. Parents and teachers

Untrue Stereotypes of Gifted People:

- Physically weak
- Socially inept
- Have narrow interests
- Emotionally unstable
- "Superhuman"
- Usually bored in school
- Do everything well
- Always make high scores on tests
- Always successful

Figure 10.2
The qualities associated with giftedness electrify our society.

What labels have been applied to you in regard to your intelligence? List a few below.

alike usually recognize students with general intellectual talent by their expansive fund of general information, productive thinking, and high levels of vocabulary, memory, abstract word knowledge, and abstract reasoning (Gifted Resources, 1996). Using a broad definition of giftedness, a school system could expect to identify 10% to 15% of its student population as gifted and talented (Gifted Resources, 1996).

Indeed, the world abounds with examples of gifted, creative people. Gifted people in politics and business include Jesse Jackson, George Bush, Bill Clinton, Bill Gates, and Michael Eisner. Many of the most outstanding minds in history were not recognized for their abilities until they were older (see Figure 10.3). For instance, Walt Disney was fired from a job because he had "no good ideas," and Winston Churchill failed sixth grade.

Figure 10.3
Physicist Albert Einstein, developer of the Theories of Relativity, failed at math as a child.

We label and use stereotypes for intellectual differences the same way we do for physical, racial, cultural, ethnic, and sexual orientation. Most of us have been called "stupid" or "idiot" at one time or another. If you have been an honor student, you have probably been called "smart," "a brain," "genius," or even some negative term like "nerd" or "know-it-all." Sometimes those who are intellectually advanced are ostracized by others because they are different. The "gifted underachiever" phenomenon refers to highly talented people with poor academic or work performance, due perhaps to boredom, alienation, the desire to fit in with peers, or the general feeling of being misplaced and misunderstood.

Developmental Delay

It is always a mistake to generalize about individuals. People with limited intellectual ability, for example, do not fit stereotypes any more than do any other group. Developmental delay is a generic expression applied to physical or mental disabilities occurring before adulthood. This label is generally preferable to "mental retardation" or "mental handicap" because it does not convey the same negative connotations. A developmental delay is not a disease. People exhibiting developmental delays do learn, but often slowly and with difficulty (see Figure 10.4). Like any other people, they have the capacity to learn, to develop, and to grow.

According to a new definition by the American Association on Mental Retardation, adopted in 1992, an individual is considered to have mental retardation based on the following three criteria: 1) an intellectual functioning level (I.Q.) below 70-75; 2) significant limitations in two or more adaptive skill areas (those daily living skills needed to live, work and play in the community, including communication, self-care, home living, social skills, leisure, health and safety, self-direction, functional academics, community use and work); and 3) the condition being present from age 18 or earlier (New Bedford Harbor, 1996).

Characteristics of people with developmental delays can include:

- Language ability below their age group
- Play interests that are immature for their age
- Difficulty in generalizing
- Poor attention span
- Poor sensory skills
- Poor motor coordination
- Low frustration level
- Poor social skills
- Poor self-concepts
- Need to hear things many times to learn them
- Need to repeat activities many times to understand them

Numerous studies have been conducted in local communities to determine the prevalence of developmental delays. The Arc (formerly the Association

Figure 10.4
People with developmental delays may learn at a slower rate.

Untrue Stereotypes of People with Mental Retardation:

- Always look different from nondisabled people
- Tend to be gentle people who make friends easily
- Should not be expected to work in the competitive job market
- Always identified in infancy
- Defined solely upon scores on an IQ test

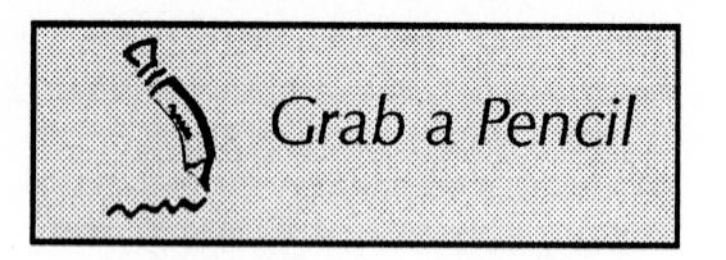

Your best friend has just called to tell you about the birth of her new baby. You are surprised to learn that the infant has developmental delays. Write a letter of support to your friend and offer any appropriate suggestions.

of Retarded Citizens) concluded in 1982 that 2.5 to 3 percent of the general population has mental retardation. A 1993 review of prevalence studies generally confirmed this distribution (New Bedford Harbor, 1996). Based on the 1990 census, an estimated 6.2 to 7.5 million people have mental retardation, which means that one out of ten American families is directly affected by mental retardation (New Bedford Harbor, 1996).

Norman Alessi (1992), an expert on developmental delays and mental retardation, suggests that emotional and behavioral disorders are a frequent complication of developmental delays and may interfere with the person's progress. Alessi points out that "Most retarded children recognize that they are behind others of their own age. Some may become frustrated, withdrawn or anxious, or act 'bad' to get the attention of other youngsters and adults" (1992, p. 1). People with developmental delays may also become depressed. Without enough language skills to talk about their feelings or emotions, their depression may express itself through new problems in behavior, eating, and sleeping (Alessi, 1992).

Causes of developmental delays fall into five basic categories: genetic, prenatal, perinatal, postnatal, and psychosocial. We will briefly examine each of these categories.

Genetic causation typically manifests itself as Down Syndrome or Phenylketonuria (PKU). Down Syndrome is caused by a chromosomal abnormality which occurs in about one of every one-thousand births. PKU is a hereditary condition involving the absence of an enzyme essential to protein ingestion, causing toxicity in the brain. PKU, if detected early enough, can be controlled by diet and therefore not cause a developmental delay.

The unborn baby is vulnerable to certain diseases and traumas. Prenatal causation has been linked to

Rubella, Syphilis, RH factor, Fetal Alcohol Syndrome, and drug addiction.

The actual birth itself is not without dangers. Perinatal causation has been traced to asphyxia, positioning, prolonged labor, and forceps delivery.

The developing child is vulnerable to accidents, environmental toxicity, and illnesses (see Figure 10.5). Thus, postnatal causation includes injuries, brain tumors, lead poisoning, and high fevers (such as meningitis or encephalitis).

Finally, psychosocial causation includes malnutrition, a poor environment, poor infant care, poor medical care, lack of stimulation, and lack of motivation. As with learning disabilities, there is a broad range of possible causes and a broad range of manifestations, some easily detectable and some not.

Figure 10.5
Postnatal causation of developmental delays includes accidents.

Researchers at New Bedford Harbor estimate that 87 percent of those with developmental delays will be mildly affected and will be only a little slower than average in learning new information and skills. As children, their mental retardation is not readily apparent and may not be identified until they enter school (New Bedford Harbor, 1996). Many persons exhibiting developmental delays graduate from high school and work to support themselves. Some live independently in group homes, and others marry and have children of their own. There is a very wide range of abilities among those who are considered to be developmentally delayed. It is important to look at the whole person and to help all individuals to accomplish their greatest potential rather than focusing on limitations.

Some have said that gifted children who are not identified and who do not receive appropriate school interventions may be our most wasted resource. Do you agree or disagree with this statement? Why?

Terminology

Following are some common terms relating to intellectual ability that you should familiarize yourself with:

Developmental delay. This refers to a limitation in one's ability to learn or care for him or herself. Once institutionalized for their condition, persons with developmental delays are now guaranteed the right to learn and grow to the best of their abilities in a more "normal" environment.

Down syndrome. A genetically linked disorder caused by a chromosomal abnormality, Down syndrome is a common cause of developmental delays.

Giftedness. Characteristics of gifted students include outstanding performance on an achievement or aptitude test in one area such as mathematics or language arts, openness to experience, setting personal standards for evaluation, ability to play with ideas, willingness to take risks, preference for complexity, tolerance for ambiguity, positive self-image, ability to adapt readily to new situations, and ability to become submerged in a task (Gifted Resources, 1996).

Gifted underachiever. This is a general term for a talented student who does not achieve academic success. Gifted underachievers may not be interested in conforming to school rules, may be disinterested or bored with coursework, and may feel alienated from their classmates.

Intelligence. This refers to a person's capacity for reasoning, understanding, acquiring knowledge, and other forms of mental activity.

Phenylketonuria (PKU). This hereditary abnormality of the metabolism can cause developmental delays.

Interactions

Following are some interaction tips for dealing effectively with people of varying intellectual abilities (adapted from Karge, 1996).

Avoid using slang terms that refer to intellectual ability. For example, "For someone so smart you're sure acting stupid," or "I've lost my car keys again—I'm so retarded."

Be consistent and fair. Do not adapt for a person who is gifted or developmentally delayed unless the person requests that.

Model self-discipline, hard work, and patience. These are valuable qualities for people of any intellectual ability.

Encourage creative problem solving. Guide people toward new ways of looking at problems.

Provide a safe environment for exploration. Encourage others to learn at their own pace and experiment in their own way.

Conclusion

What would you consider "average intelligence" to mean? Remember as you try to define "average intelligence" that according to U.S. government studies approximately one in four Americans reads below a fourth grade level. Intelligence does not mean only education—consider life

I remember one day in school. It was a long time ago but I remember. Someone painted the words 'Wip Room' on the door to the class. Our teacher was real mad. Some of the girls was crying. That's because the other kids always teased us cause we was retarded. I used to be late lots so the other kids wouldn't see me going in there. But I learned a lot. And now I got a good job. I work on the 30th floor. I wrap up packages. And I deliver them all around the city. I use the subway to deliver the packages. Sometimes I have to get them there real fast. I like my job and where I live. I like my friends at the group home. And I like spending the money I make at my job. And I like to go to movies.
—R.W., New York City

skills, talents, and mechanical aptitude as well. Persons who are intellectually superior or who have significant intellectual deficiencies are contributing members of society. They deserve our respect in the same way that others who are diverse do.

Labels often become self-fulfilling prophecies. Rosenthal and Jacobson (1968) demonstrated that to some extent, children's performance can improve if teachers, relatives, and friends are led to believe they have superior ability. For example, students labeled as gifted stereotypically excel at most tasks, while students labeled as developmentally delayed stereotypically perform poorly. Our students actually learn to perform and conform to what is expected of them. When it comes to intellect, we should think of all people as organic computers—waiting to be turned on, given a purpose, and allowed the freedom to operate on their own.

References

Alessi, N. (Ed.). (1992). Children who are mentally retarded. In *Facts for families*. [On-line.] Available: http://www.aacap.org/factsFam/retarded.htm

Gifted Resources. (1996). A short summary of giftedness. In *Gifted resources homepage*. [On-line.] Available: http://www.eskimo.com/~user/zbrief.html

Karge, B. D. (1996). Intellectual differences. In S. E. Schwartz & B. D. Karge, *Human diversity: A guide for understanding* (2nd ed.) (pp. 188-189). New York: McGraw-Hill.

New Bedford Harbor. (1996). Introduction to mental retardation. In *New Bedford Harbor services*. [On-line]. Available: http://www.ici.net/cust_pages/nbhsi/mr-def.htm

Rosenthal, R. & Jacobson, L. (1968). *Pygmalion in the classroom*. New York: Holt, Rinehard & Winston.

Suggested Readings

Armstrong, T. (1993). *Seven kinds of smart: Identifying and developing your many intelligences.* New York: Plume Books.

This book discusses how to identify and develop many different types of intelligences.

Rimm, S. B. (1986). *Underachievement syndrome: causes and cures.* Watertown, WI: Apple.

The author believes underachievement syndrome is epidemic and discusses how parents and teachers can best work with persons who are underachieving.

University of Connecticut. (1996). Overview of mental retardation. In *Department of mental retardation homepage.* [On-line.] Available: http://www2.uconn.edu/ctstate/dmr/overview.html

This is a valuable resource on current knowledge of developmental delays.

Notes

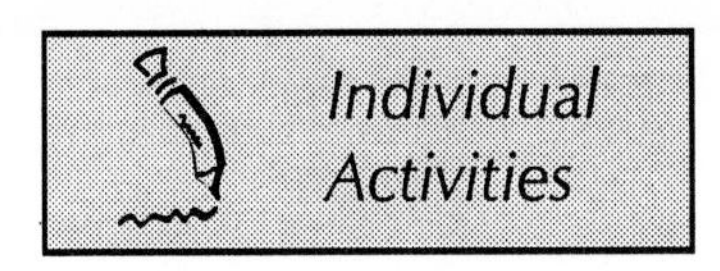

1. Have you ever tried to complete a task or participate in an activity and found it beyond your grasp (such as playing a sport you were not familiar with, trying to play a musical instrument, or beginning a class in a hard subject)? This is a very awkward feeling and, unfortunately, how most people with mental retardation feel every day. They have feelings just like you. List 10 feelings that you have felt when placed in uncomfortable situations due to your lack of ability.

____________________ ____________________

____________________ ____________________

____________________ ____________________

____________________ ____________________

____________________ ____________________

Can you now relate to the feelings of a person with mental retardation? Put a check in front of those items which are probably similar for people with mental retardation.

2. When you think of gifted people, what descriptors or labels cross your mind? List the first 10 thoughts that come to mind. Then check the ones which are accurate labels for gifted people.

____________________ ____________________

____________________ ____________________

____________________ ____________________

____________________ ____________________

____________________ ____________________

Group Activities

1. Divide up into groups of five or six students. Your assignment is to brag. Have each person describe his or her special talents. You will probably be surprised to learn how many of your fellow students are gifted in specific areas.

Reactions:

2. It is highly probable that many people in your class know someone or have a family member who is developmentally delayed. Ask those classmates to tell everyone about that person's abilities, limitations, and personality.

Reactions:

Reflection Paper 10.1

What do you think makes an individual gifted? Is giftedness an inherited quality or does one come by it through experience and hard work?

Reflection Paper 10.2

There are many misconceptions concerning what people with developmental delays can and cannot do. From your experiences and knowledge, discuss three of these misconceptions and what the truth actually is.

Notes

Chapter 11 Health Differences

Chapter Sections

- Introduction
- Types of Health Impairments
- HIV and AIDS
- Drug and Alcohol Abuse
- Other Major Health Concerns
- Additional Terminology
- Promoting Vitality
- Support Services
- Interactions
- Conclusion
- Individual and Group Activities

"If you don't have your health, you don't have anything."
—Anonymous

Objectives

- Distinguish between acute and chronic health impairments.
- Identify where a person can look for help in attempting to cope with health difficulties.
- Explain six ways to boost one's immunity.
- Identify five ways to foster healthy interactions with people undergoing an illness.

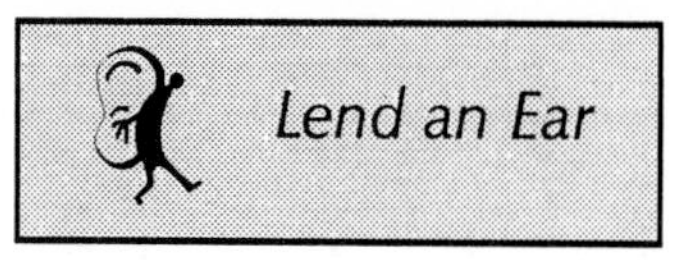

"Nobody can be in good health if he does not have all the time fresh air, sunshine, and good water."
—Sioux Chief Flying Hawk

Introduction

Health means *vitality*—the energy of life. To be in good health means to feel strong, be alert, and function normally in all ways. Health is not just lack of disease. Health also means enjoying life and feeling successful in what we do. Every day we are exposed to millions of germs and diseases, but healthy immune cells are strong enough to fight those diseases. When the immune system gets run down from stress, one might get sick enough to interfere with normal activities and perhaps even require hospitalization. In addition, there are traumas from such things as physical assaults, sports injuries, and accidents involving automobiles, bicycles, skateboards, and boats. Although some illnesses have no cure, medicine, caring, and other treatments can help to improve a person's quality of life. Clearly, health impairments have a significant impact upon society. Knowledge and understanding of health issues can help us to offer sensitive and appropriate assistance to others (Krementz, 1989). This chapter will also show how you can make a real difference by promoting well-being (see Figure 11.1).

Figure 11.1
Loving relationships are an important way of promoting health and well-being.

Maintaining a good quality of life is important for people who have a permanent health disability and for members of their family. What do you consider a good quality of life to be? What can be done toward maintaining a good quality of life?

Types of Health Impairments

Individuals with health impairments usually do not appear to be any different from other people, and they can engage in most typical activities. During periods of good health there may be no disease symptoms at all. Health impairments include limited strength, vitality, or alertness. Such impairments are frequently not noticeable to the general public unless an acute episode occurs. Acute impairments are time-limited, while chronic disorders are considered to be treatable but not curable. Such things as diabetes are chronic, while injuries suffered in a fall might be acute.

Impairments may occur as the result of conditions present at birth or may result from diseasse. No two health challenges are the same, nor do they affect people in the same way. Each disease has individual causes, symptoms, cures, and interventions. It is critical to support those around us in a loving manner. Treating people the same as we treat others and communicating with them about their health problems are two key ways we can improve their well-being without offering any medical or psychological advice.

HIV and AIDS

A serious health issue which all of society faces is Acquired Immune Deficiency Syndrome (AIDS). AIDS is a life-threatening disease for which there is as yet no cure, but it can be prevented. AIDS education has become the responsibility of not only parents, teachers, and medical personnel, but of every thinking individual.

If you are sexually active (whether through heterosexual, homosexual, or bisexual contact), have shared intravenous needles, have received a blood transfusion prior to 1985, or have been involved in any activities where blood or bodily fluids are transmitted, then you are at risk of becoming infected. If this description fits your sexual partner, then you are also at risk. Human Immunodeficiency Virus (HIV) has been linked to the onset of AIDS. Not everyone infected with HIV gets AIDS. When one is infected with HIV, his or her immune system attempts to fight the virus but is unable to destroy it.

AIDS is a serious condition that destroys the body's natural defenses against disease. A physician will diagnose an HIV-positive individual with AIDS if one of the serious opportunistic infections associated with being positive develops. These opportunistic infections cause malignancy, serious weight loss, AIDS-related dementia, or a T-cell count drop to below 200 (versus a normal count of between 900 and 1500 T-cells).

The spread of AIDS can be stopped through awareness, compassion, commitment, and community support for safer sexual activities. Most medical professionals consider the following activities "safe":

- Abstinence from sexual activities
- Casual contact
- Touching
- Hugging
- Dry kissing
- Fantasizing
- Masturbating

Professionals further point out that AIDS is not spread through such activities as:

- Coughing
- Sneezing
- Breathing the same air
- Eating together
- Eating food prepared by a person with HIV
- Donating blood
- Working or socializing with someone with HIV

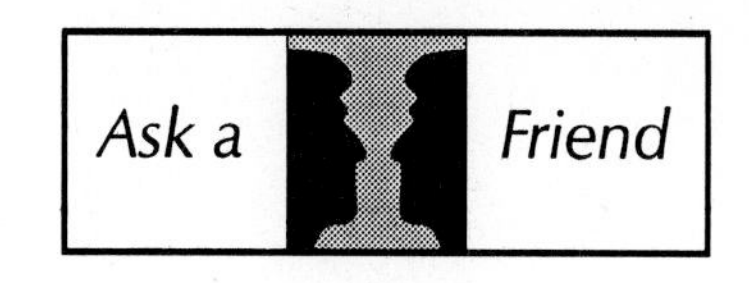

When it comes to HIV and AIDS, exactly what activities are considered "safe"?

HIV is spread by blood-to-blood contact, sharing bodily fluids during sexual contact, and breast milk. A pregnant woman may be able to pass HIV to her baby before or during birth. Caution should always be taken. If you have questions, contact your doctor or the AIDS Foundation.

The most common way to pass HIV is through sexual activity. The following activities are presumed *risky:*

- Deep (french) kissing
- Vaginal/anal intercourse
- Fellatio (mouth/penis contact)
- Cunnilingus (mouth/vaginal contact)
- Analingus (mouth/anal contact)
- Sharing non-sterile needles from drugs, steroids, tattooing, or body piercing

These activities can be additionally hazardous if performed under the influence of alcohol and/or other substances.

The only 100% "safe-sex" is abstinence. Protection devices are never guaranteed. Condom usage is strongly recommended by AIDS researchers. When used correctly, condoms collect semen that is discharged before and during ejaculation while acting as a barrier to prevent sperm from entering a partner's body. Latex condoms containing a spermicide to kill sperm and sexually transmitted diseases are recommended (see Figure 11.2).

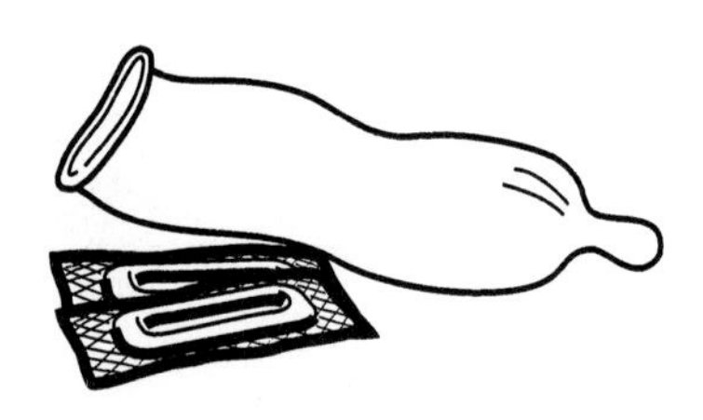

Figure 11.2
Though not 100% effective, condoms are recommended to help reduce the risk of HIV transmission.

Symptoms commonly associated with HIV infection are persistent and unexplained swollen lymph glands, weight loss, tiredness, loss of appetite, night sweats, diarrhea, fever, and mental disorders. If these symptoms continue for more than two weeks, a person should see a physician. None of these symptoms individually means a person has AIDS.

It is critical that an HIV antibody test be performed if AIDS is suspected. The test determines the presence of HIV antibodies in the blood. If there are antibodies present, the person is infected with HIV. Though home tests are now available which ensure privacy and confidentiality, counseling sessions are recommended both at the time of the test and when the results of the test are known.

If the diagnosis is HIV positive, early medical help is one way to remain healthy. It is vital for those who are HIV positive (or at risk for HIV) to strengthen their immune systems. A proper diet, regular exercise, stress-reducing programs (such as meditation or other relaxation techniques), abstinence from drugs and alcohol, regular massage therapy, and loving relationships have all been shown to dramatically boost immunity and general well-being. Medical research into the treatment and cure of AIDS is ongoing, and new drug therapies have been quite promising.

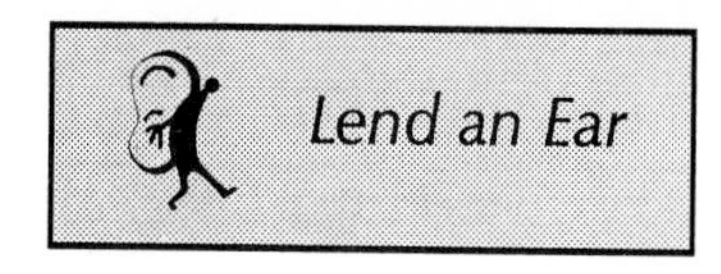

"Healing is a matter of time, but it is sometimes also a matter of opportunity."
—Hippocrates, "Father of Medicine"

Drug and Alcohol Abuse

Another major health concern which affects growing numbers of people, including children of increasingly younger ages, is alcohol and drug abuse. An estimated 14 million Americans abuse alcohol or illicit drugs to the point of dependence each year. Pregnant mothers who abuse drugs place themselves as well as their babies at risk for a variety of serious and sometimes life-threatening problems which affect the children into adulthood. Seizures, shortness of breath, lung damage, nasal membrane burns, respiratory paralysis (by overdose), cardiovascular problems, anorexia, and premature labor are all examples of potential problems. Alcohol and drug abuse have other negative effects, including the increased risk of impaired physical and intellectual well-being and accidental injury of oneselves and others (see Figure 11.3).

Figure 11.3
Alcohol abuse leads to a variety of life-threatening problems.

Besides alcohol, a commonly used substance which can be legally purchased by adults and is widely available to children is tobacco. Snuff, chewing, or smoking tobacco products are popular sources of the drug nicotine. Although the number of American men who smoke is decreasing steadily, the percentage of female smokers has been rising, as has use by children and teens, according to recent studies. As a result, today more women die of lung cancer than breast cancer each year. The use of snuff or chewing tobacco has increased in the U.S., and tobacco smoking in other parts of the world remains prevalent. Recent cancer studies have proven the dangers of secondhand smoke, and most government and public establishments—such as restaurants, schools, hospitals, and airplanes—are now designating themselves "smoke-free zones" to prevent an unhealthy environment . All cancers caused by smoking could be prevented entirely (see Figure 11.4). Nicotine is highly addictive, and people in the process of breaking the habit may exhibit irritability, anxiety, cravings, headaches, and lethargy. Symptoms of withdrawal may continue for 4-8 weeks and cravings may continue indefinitely.

Figure 11.4
All diseases caused by smoking, including lung cancer, could be entirely prevented.

Other Major Health Concerns

Now let's take a brief look at some other major categories of health impairments you may encounter.

Asthma is a chronic respiratory condition in which breathing becomes difficult due to blocked air passages. It affects almost 10 percent of children in the United States (Bachman, 1992), many of whom carry the symptoms into adulthood. Episodes of asthma can be triggered by emotional factors, which may tighten the muscles around the bronchial tubes and cause swelling of the tissues. Many new drug treatments successfully control the symptoms of asthma (see Figure 11.5). People with asthma may need to keep inhalers or

Figure 11.5
Inhalers send medicine directly to the lungs of a person suffering an asthma attack.

oxygen with them, especially when there are allergens such as pets or plants nearby (Bachman, 1992). Asthma attacks vary from person to person, as people react to different triggers.

What are the proper steps in handling a seizure?

Diabetes is a metabolic disorder characterized by the inability to properly process carbohydrates. In other words, it affects the means by which the body changes the food we eat into energy. Typically, the pancreas fails to secrete an adequate insulin supply, resulting in an abnormal concentration of blood sugar in the blood and urine. Symptoms include excessive thirst, frequent urination, weight loss, slow healing of cuts and bruises, pain in joints, and drowsiness. Long-term problems could include blindness, kidney failure, and heart attacks. Individuals with diabetes may need to snack regularly, eat a piece of candy, or take a glucose tablet to regulate their blood sugar. Many persons with diabetes take daily doses of insulin either orally or through self-administered injections.

Epilepsy is a chronic central nervous system condition characterized by periodic seizures, convulsions of the muscles, and sometimes a loss of consciousness. A major epileptic seizure is often dramatic and frightening to those who have little experience. Typically, a seizure lasts only a few minutes and does not require expert care. Epilepsy can usually be controlled by medication, and new medical procedures appear to be promising. In case of a seizure, the person should not be moved unless there is danger of an injury from banging his or her head against on object. Do not put an object in the person's mouth. Loosen any tight clothing. Turn the person on his or her side so that no mucus or blood is inhaled. If the seizure lasts longer than five minutes, call an ambulance (Reisner, 1988).

There are many other severe illnesses which may affect people around you. These may include such

I was 10 years old when my mother was diagnosed with leukemia. It was as if we were all diagnosed. My brother's, my sister's, and my life were changed forever. We had the disease, too, in a way, because we had to deal with all the symptoms.

Mom was hit unusually suddenly, and she went from healthy to unable to walk or feed herself in a matter of weeks. It took everybody pulling together to cope with the added responsibilities and medical expenses. We cried and cursed a lot, but we all pitched in.

We learned a lot from Mom's strength and also learned a lot about our own ability to do what was required. I missed a lot of school that first year, but my teachers were very understanding and helped me to make up the work. They also taught me and my classmates about the disease.

We actually got stronger individually and as a family in spite of all the hurt and worry. People don't appreciate their health until they lose it.

—William R., Springdale, PA

diseases as cancer, hepatitis, sickle-cell anemia, and severe allergies. Should you meet people with these or any other health problems, it would be wise for you to conduct your own research into these conditions and discuss the specific cases with your health care provider. This will ensure that your interactions with these people are proper and helpful.

Additional Terminology

Knowledge of the following additional health impairments is vital for all educated members of society.

Hemophilia. This is a disease characterized by the blood's failure to clot after injury as well as profuse bleeding from even minor injuries. It is hereditary, found primarily in males because females carry the hemophiliac gene, passing it to male children.

Leukemia. This is cancer of the blood-forming organs. It results in an increase of white blood cells and progressive deterioration of the body.

Rheumatic Fever. Characterized by acute inflammation of the joints, fever, nosebleeds, rashes, and nervous disorders, this can cause heart damage by scaring tissue and valves. It often appears after a streptococcus infection.

Sickle-cell anemia. This is a blood condition in which the red cells assume a sickle shape and impair circulation by not properly carrying oxygen. No cure is available for this condition. Any activity that reduces oxygen in the blood (such as hiking to high altitudes) may bring about a crisis. The symptoms are low vitality, pain, shedding of blood cells, interference with cerebral nutrition, and chronic illness. The condition is genetic and largely limited to persons of African descent.

Tuberculosis. This is an infectious, chronic and communicable disease. It most often affects lungs, but also destroys tissues of body organs and bones. A positive tuberculin skin test reaction can diagnose the disease. Properly treated, patients are cured.

Assume that you just read a letter to the editor of your local newspaper which suggested that persons with AIDS should not be allowed to shop in your community's food stores. Write your own letter to the editor in response.

Promoting Vitality

You don't have to be a doctor to promote vitality. Challenge yourself to make a difference in other people's health and well-being every day by doing the following:

- smile
- show appreciation
- show understanding
- be considerate
- give a hug
- lend a hand
- lend an ear
- offer a shoulder
- express your feelings
- pay a compliment
- offer good wishes

Giving someone your time, support, encouragement, and love can ease the stresses of being ill (see Figure 11.6). You can also do things to promote your own vitality. Challenge yourself to do the following things every day:

- experience nature
- slow down
- enjoy quiet time
- listen to your body
- eat nourishing foods
- have loving relationships
- abstain from drugs and alcohol

Figure 11.6
Expressing love for someone can significantly reduce the stresses of illness.

Support Services

Breakthroughs in medical technology extend life but often necessitate lengthy and costly treatments and hospital stays, often resulting in psychological stresses for both the individual and his or her family (see Figure 11.7). For people attempting to cope with such difficulties, advocacy and support groups are available for almost all categories of health issues. If people with health impairments or their families ask for your advice or are in obvious distress, you might encourage them to look for a network of people interested in the same issue who can help them learn to cope with the associated challenges.

Figure 11.7
Lengthy hospital stays may add psychological stresses to being ill.

Interactions

Never judge by appearances. Many seriously ill people look healthy.

Keep a current certification in first aid and CPR. Be well-practiced in emergency techniques.

Caring, understanding, attention, and touch are some of nature's most potent medicines. Give your time, support, and encouragement and you will ease the stresses of being ill.

Know the medications a person must take. See that he or she has the opportunity to carry out the prescription.

Be inclusive rather than exclusive. Involve the person in all activities that you would normally, depending upon his or her interests and abilities.

Talk about coping with the health challenge. Since many health conditions are controlled rather than cured, it is appropriate to talk about coping with the ongoing symptoms. Nurture and promote positive coping mechanisms.

Listen to the person's concerns. Sometimes simply listening and hearing someone's feelings can be helpful to all of you.

Learn about the health-related services available in your community. Be prepared to access them as needed.

Seek an understanding of how it feels to both the person and his or her family to face a health challenge. With learning comes growth and healing.

Act deliberately and calmly at all times. Your appropriate behavior will serve as a model to others, and help to allay their anxiety.

Conclusion

In response to rising health care costs, current trends in healing are focusing more and more on prevention and holistic health, which treats the entire person and his or her lifestyle, not merely the disease itself. Today, hospitals and doctors' offices are not the only places one can turn to for help. People with health issues are finding guidance, support, and treatment from dieticians, massage therapists, chiropractors, acupuncturists, herbalists, psychiatrists, physical therapists, exercise and fitness counselors, families, and friends.

The health and safety of those around you is a major responsibility. Most unexpected events that will involve you will be fairly minor and should be handled logically and easily. Some events require extreme care and immediate response. Your greatest challenge, however, may be to keep in mind that people who suffer chronic illness, disabilities, or acute medical problems have often experienced more difficult situations, endured more pain and discomfort, and made more significant adjustments in their lives than many people will ever be called upon to do. Approach them as you would anyone, with sensitivity and respect for the individuals that they are.

References

Bachman, J. L. (1992). *Keys to dealing with childhood allergies.* Hauppauge, NY: Barrons.

Krementz, J. (1989). *How it feels to fight for your life: The inspiring stories of fourteen children who are living with chronic illness.* New York: Fireside Books.

Reisner, H. (Ed.) (1988). *Children with epilepsy: A parents guide.* Rockville, MD: Woodbine House.

Suggested Readings

Cain, N. (1996). *Healing the child: A mother's story.* New York: Rawson Associates.

This is the personal account of a mother who undertook the daily health care of her child.

Elliott, J. (1990). *If your child has diabetes.* New York: Perigee Books.

This book provides information for parents and families dealing with diabetes.

Jennings, C. (1993). *Understanding and preventing AIDS: A book for everyone.* Cambridge, MA: Health Alert Press.

This is an essential guide for anyone interested in gaining knowledge about AIDS.

Individual Activities

1. When you think of diabetes, asthma, epilepsy, and AIDS, there are several personal problems that may go along with these conditions. Describe a few problems associated with each condition. Suggest strategies to overcome these problems.

2. How would it feel not to eat your favorite foods, play sports, or have to take medication or injections on a daily basis? These are considerations that people with health impairments may face every day. List 10 thoughts that are going through your mind while you think about this.

Group Activities

1. How would you feel if a member of your family had a severe health impairment? With four or five other students, discuss ways you would deal with the situation.

Reactions:

2. Arrange a class visit to a nursing home which specializes in caring for patients with severe health impairments. While there, be sure to notice the level of care that is required for some patients. Try to focus on the attitudes of both staff members and patients.

Reactions:

Reflection Paper 11.1

Your local school board has voted to have all students with HIV taught in one school building so they would not be interacting with other students. What is your reaction to this ruling?

Reflection Paper 11.2

What guidelines would you suggest for interacting with people who are health impaired? Use examples of different types of health problems in your response.

Notes

Chapter 12 Communication Differences

Chapter Sections

"Communication is the back and forth of telling and listening and responding, so you know you are not alone."
—A. Brandenberg, author

Objectives

- Identify four ways people communciate, besides speech.
- Identify six common communication disorders.
- Explain the best way to communicate with someone who is not proficient in English.
- Explain the best way to handle speech and language disorders.

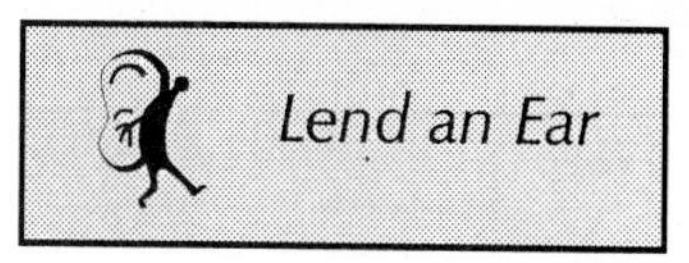

"The mystery of language was revealed to me. I knew then that 'w-a-t-e-r' meant the wonderful cool something that was flowing over my hand. That living word awakened my soul, gave it light, set it free."
—Helen Keller, author and educator

Introduction

Communication is a relationship—a human connection. The word *communicate* comes from a Latin word which means "to share." Communication is a means of transmitting information, ideas, or feelings from one person to another (see Figure 12.1). The ability to communicate often means survival, as when we need to signal our need for help.

Communication is clearly the central cohesive element of society. Through communication we share our knowledge, ideas, dreams, and hopes. We communicate in many ways, including speech, gestures, drawings, written language, sign language, and facial expressions. Language is the primary means by which people communicate. Ironically, however, it is also the primary means by which people *fail* to communicate. Anyone who has traveled has no doubt encountered either an unfamiliar dialect or a totally foreign language. In different cultures, people communicate with different gestures as well. The purpose of this chapter is to look at the diverse ways in which people exchange ideas and to examine methods to break down communication barriers.

Figure 12.1
Communication allows individuals to share their ideas and feelings with others.

The Culture of Communication

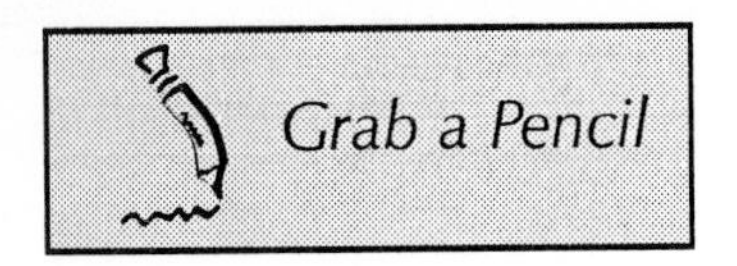

Over the generations, our society has passed down bits of folk wisdom which we sometimes call "old sayings." We often hear these proverbs from our grandparents. For example, "Absence makes the heart grow fonder." What are your favorite old sayings, and how do they help to express and preserve your culture?

Our reliance on language is so fundamental that we often take it for granted. It is through language that we organize our social structures and learn how to coexist. The communication of messages plays a vital role in expressing and transmitting our culture. In fact, culture itself has been defined as "any system in which messages cultivate and regulate relationships" (Gerbner, 1990, p. 423). Whenever you speak and write your language, you also speak and write your culture. The cultural assumptions and understandings of many generations are embedded in your words.

A culture's myths, legends, and folk ballads are products of communication. Before our ancient ancestors invented writing, they preserved their history through the "oral tradition." They recited or sang their history over and over again, thereby passing it down to the next generation. Over time, the stories changed and kept only the most important and profound meanings—ideas about the beginnings of life, the nature of humanity, and the mysteries of death. These stories survive today as mythology, folklore, legends, allegories, fables, and parables.

A culture's visual arts are another important product of communication. The earliest records of human communication are the prehistoric drawings on cave walls in France and Spain, depicting human figures and animals. To this day, art continues to serve as an important visual means of sharing information and imagination, whether in the form of a painting in a museum or an illustration in a magazine advertisement. Written language, too, can reach artistic proportions in poetry and literature. Even the alphabet itself can be a means of artful expression, as seen in calligraphic handwriting (see Figure 12.2).

Figure 12.2
The written word itself can be an artform, as in Chinese calligraphy.

There are yet other facets of cultural communication found in the performing arts. Dance, for example, is a system of body language that expresses meaning through movement, gestures, poses, and pantomime. In many parts of the world, such as India, Japan, and Thailand, complex systems of pantomime and dance are combined with symbolic hand gestures, facial expressions, and body movements to tell a story (Crystal, 1987). Some of these dances have survived for thousands of years and communicate traditions and ideas from the past to every new generation. Similarly, opera combines singing, costumes, movement, and other modes of communication. Some people enjoy going to the opera even if they don't understand the words of the songs because they can understand the story through the universal language of gestures, facial expressions, and vocal intonations.

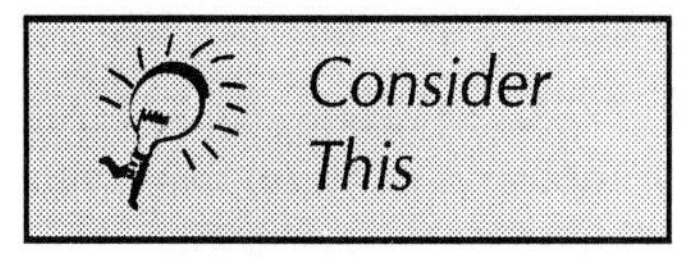

Consider the ways in which the ability to communicate means survival, both from a daily, personal perspective and from a worldwide, humanitarian perspective.

Body Language

Body language is a form of nonverbal communication. Through posture, facial expressions, eye contact, and hand gestures, we can express our feelings, emotions, and attitudes. Waving a forefinger back and forth can say "You are wrong." Holding someone's gaze for a number of moments can indicate "I know you" or "I am interested in you." Leaning forward during a conversation can mean "I am involved in what you are saying." Some body language is universal. For example, certain facial expressions or gestures signify pain, fear, and joy in every culture (see Figure 12.3). However, other expressions

Figure 12.3
Facial expressions are one popular means of communicating.

may have different meanings in different cultures. Nodding one's head up and down, for instance, means "yes" in the United States and Europe but may mean "no" in other cultures (Beier, 1990).

In some cultures, touching is a very important part of communication. Touching can involve a wide variety of activities: embracing, holding hands, kissing, linking arms, nudging, patting, shaking hands, and slapping, just to name a few. Tactile activities express basic social interactions such as gratitude, greeting, leave taking, congratulation, sexual interest, aggression, and affection. Some societies are more tolerant of touching than others, depending upon concepts about personal space (see Figure 12.4). Northern Europeans and Indians, for example, tend to avoid touching, while Arabs and Latin Americans tend to favor it. In a study of couples sitting together in cafés, it was found that Puerto Ricans touched each other 180 times an hour while Londoners never touched at all (Crystal, 1987). This does not mean that Puerto Ricans are more affectionate than Londoners, just that in British culture communication is less tactile.

There is a science of "visible speech" called *eurhythmy,* in which the body symbolically interprets the sounds of language. Each sound that a person

Figure 12.4
Personal space refers to an individual's physical comfort around others. Personal space can be divided into four basic zones. (Figure derived from Conley, 1997.)

1. PRIVATE ZONE
Extending from physical contact to 18 inches from the body, this zone is for intimate, personal interactions.

2. FRIENDLY ZONE
Extending from 18 inches to 4 feet from the body, this zone is for interacting with people we know well.

3. SOCIAL ZONE
Extending from 4 to 12 feet from the body, this zone is for casual interactions.

4. PUBLIC ZONE
Extending from 12 to 25 feet from the body, this is for impersonal interactions.

can articulate is reflected by a body movement. The sound *a*, for example, means astonishment and wonder and is shown by raising the arms over one's head as if holding a giant ball. The sound *u* means something is chilling and is expressed by pressing the arms and legs together. Sound, movement, and meaning come together in this intricate system of body language.

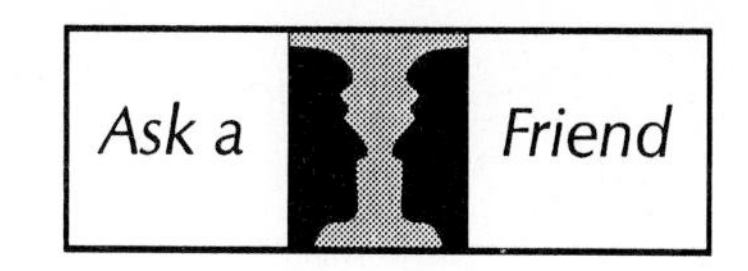

What are three good reasons to learn another language?

Written and Spoken Language

There are between 4,000 and 5,000 languages spoken around the world. Mandarin predominates, with nearly one billion speakers, and English comes in a distant second with less than half the number of speakers (see FIgure 12.5). David Crystal (1987) has explained that all languages are "equal in the sense that there is nothing intrinsically limiting, demeaning, or handicapping about any of them" (p. 6). No one language is necessarily easier or more difficult to learn or speak than another—even the languages of primitive societies have complex grammatical rules.

Because every language meets the social and psychological needs of its speakers, the study of languages can provide you with valuable insights into human nature and society (Crystal 1987). That's because a society's intellectual heritage and cultural traditions are directly shaped by the society's language. If you become fluent in another language, you will be able to read great books in the original language of the author and meet the great minds of another culture on their *own* terms. You will learn how other people think, and then you will truly understand another culture and appreciate its values. Ultimately, your own heritage will be enriched because you will see it from an outside perspective. If your native language is English, or if you are in the process of learning English, you will discover that it has borrowed from many other languages.

Figure 12.5

Principal Languages of the World

Language	in millions
Chinese	999
English	487
Hindi	457
Spanish	401
Russian	280
Arabic	230
Bengali	204
Malay-Indonesian	164
Portuguese	186
French	126
Japanese	126
German	124
Urdu	104

(Famighetti, 1996)

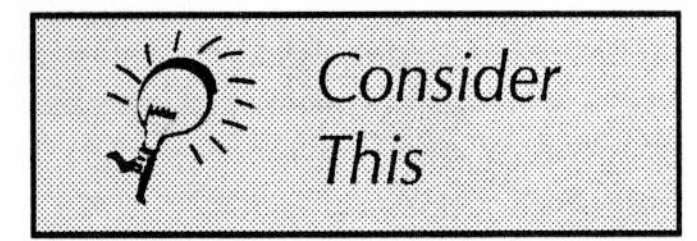

A free Esperanto course is available on the Internet at http://wwwtios.cs.utwente.nl/esperanto/hypercourse/index.html.

There is as of yet no such thing as an international language. Through the Middle Ages, Latin was the language of education in western Europe. From the 17th to the 20th century, French was the international language of diplomacy. Though today there are more Chinese speakers than any other, the complexity of the Chinese writing system discourages its use. English has assumed special status internationally, though it is still not a world language.

Attempts have been made to design an artificial language—a new, simplified system combining elements of natural languages—to serve as an international language. The artificial language Esperanto, invented in 1887 expressly to facilitate global communication, has millions of speakers worldwide. For over a hundred years, it has proven to be a genuinely living language, capable of expressing all facets of human thought (Belinfante, 1997). Esperanto belongs to no single nation or people; rather, it acts as a bridge among cultures. It has a vast literature, with both translated and original works on countless subjects, as well as periodicals and regular radio broadcasts. Esperanto's simple and very flexible structure, with a vocabulary of international character, allows reaching fluency up to ten times faster than in any other language (Belinfante, 1997). The World Congress of Esperanto, the International Youth Congress, and other meetings take place in different countries every year, uniting Esperanto speakers (see Figure 12.6). The United Nations has yet to grant Esperanto international status, however.

Figure 12.6
Esperanto is a living, international language.

Hello, friend =
Saluton, amiko!

Children's Development of Language and Speech

The age at which children develop language varies greatly. Generally speaking, however, we can safely say that the earliest communication begins with social interactions between the caregiver and the baby. For example, the caregiver may play peek-a-boo with the baby, point to and name objects, make facial expressions, and so on.

During the first half-dozen months, babies typically experiment with vocalizing simple sounds. This "cooing" gradually becomes a string of babbling, and at about age one the child may put different syllables together and may have learned to "answer" when spoken to.

Between ten and eighteen months, children usually say their first words. At this stage they often echo what they hear without understanding the word or mispronounce a word and therefore remain misunderstood. At about a year and a half, they typically make their first two-word combinations, such as "more juice."

At two years, children begin to effectively communicate with simple sentences and a vocabulary of several hundred words. After age two, the child's vocabulary and grasp of the language continues to grow. However, because the development of language varies so greatly from child to child, it is often difficult to properly diagnose communication disorders at an early age.

First Person

I've lived in the U.S. since I was five, and my sister was born here. I'm not a U.S. citizen, I'm Chinese, but I've grown up like an American and gone to school here all my life. The only Chinese people I know are my parents. They speak Mandarin at home and live as much a traditional Chinese life as they can. It's an advantage to understand two languages and cultures, but it's hard, too, because culturally I'm an American.

My parents are often shocked and disturbed over things my friends' parents wouldn't even notice. For instance, a female classmate sent me a Christmas card last year. My mother saw that the girl had written "I hope all your hopes and dreams come true," and she got hysterical. She thought that meant that we were having sex.

—Lee P., Atlanta, GA

Speech and Language Disorders

Several million people in the world are unable to communicate effectively with others because of speech or language problems. The National Institute of Health estimates that over fourteen million Americans alone experience such difficulties (Voice Foundation, 1994). The problem is magnified immeasurably because anyone who tries to communicate with a person with a speech or language disorder may also experience difficulties. Speech and language disorders inevitably draw attention to themselves, and the awareness itself may inhibit communication. The challenge, then, is to overcome

Untrue Stereotypes Regarding Communication Disorders

- People with communication disorders tend to also be mentally retarded
- Stuttering is an indication of an extremely high IQ
- Articulation disorders are easy to correct
- A person with a cleft palate always has difficulty speaking
- People with communication disorders have emotional disorders as well
- A language disorder by definition means difficulty with speech

this barrier, not to create yet a further problem with communication. With information and understanding, we can avoid this pitfall.

Speech and language disorders are actually quite different. An individual with a speech disability usually has some level of difficulty in communicating because of such characteristics as an incorrect pronunciation of words or parts of words, or a lack of fluency when saying words or phrases.

While the person with a speech disorder may have problems expressing him or herself, the person with a language disability may have either an expressive or a receptive difficulty. When an individual has an expressive language problem, that person is able to correctly produce the sounds necessary for speech but the words themselves may be used incorrectly or illogically. A person who has a receptive disorder has no known hearing difficulties and usually has knowledge of the meaning of words. However, the individual may have difficulty comprehending what has been correctly heard.

Some other common communication problems include:

Aphasia. This condition is usually caused by brain damage and characterized by labored speech and an inability to choose the right words.

Articulation problems. A difficulty with pronunciation or a lisp is considered an articulation problem. individuals with articulation problems may experience anxiety and embarrassment, which interferes with communication.

Cleft lip and palate. A cleft lip is a congenital splitting of the upper lip. A cleft palate is a congenital fissure along the middle of the palate. Both conditions may affect the development of speech, but not in all cases.

Dyslexia. This is an inability to read and spell correctly, despite normal intelligence. Frequently, letters become reversed or out of order.

Fluency disorders (stuttering). This disorder of fluency affects one's ability to control the rhythm and timing of speech. Stuttering often involves a repetition of sounds, syllables, words, or phrases. Though people who stutter may speak with difficulty, this has nothing to do with their intelligence. People who stutter frequently experience anxiety and embarrassment, and communication is invariably affected.

Illiteracy. The inability to read and write is called *illiteracy*. One in five Americans is functionally illiterate, meaning they do not have the reading or writing skills required to function effectively in society. Often illiteracy is the result of another disorder such as vision impairment, hearing impairment, or dyslexia.

Laryngeal abnormalities. Malignant growths in the throat may require surgical removal of the larynx. Persons who have undergone an laryngectomy must learn to speak by vibrating their esophagus or by using an artificial larynx which, when placed against the neck while they are talking, emits a speech substitute which is often characterized as having a "buzzing" sound.

Muteness. This is the inability to speak, due either to a physical abnormality or to emotional stress (as when someone "loses her voice").

Vocal problems. Raspy, hoarse, nasal, breathy, weak, or abnormally loud voices may make someone difficult to understand. Disorders of vocal expression are usually due to an anatomical abnormality in the vocal tract, such as the formation of nodules or polyps.

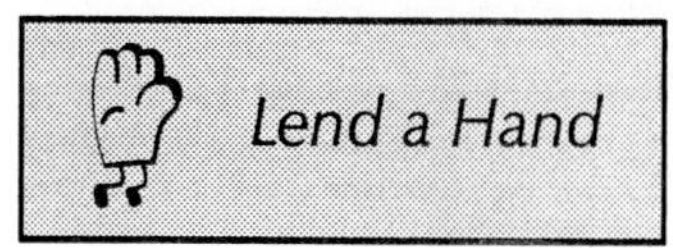

Volunteer an hour a week at your local adult literacy center to help someone learn to read.

Issues Persons with Communication Disorders May Face

- Ridicule from others
- Classroom problems
- Finding support
- Finding role models and mentors
- Relationships
- Labeling
- Finding the "right" type of treatment
- Finding acceptance
- Getting a job that accepts or is willing to recognize communication challenges
- Others considering them either less or more intelligent

Recall a situation in which you were unable to effectively communicate with someone. What was the problem and how did you (or could you have) overcome it?

Support Services

Local chapters of service organizations, such as the American Speech-Language-Hearing Association, may be able to provide information on communication disorders and educational programs. Speech and language therapists in private practice and in hospitals may be able to provide information on scientific research and methods of treatment. Speech and language therapists and local organizations also may be able to refer an individual to a support group which meets regularly.

Interactions

Language differences. It is sometimes necessary to be patient and understanding when speaking with someone from another country. It may help your patience to imagine yourself speaking someone else's native tongue. It is difficult to hear and speak a foreign language, and people from other countries are not always proficient in English. They may not be used to making the foreign sounds of our language. Mistakes in pronunciation or grammar do not mean that they are uneducated or unintelligent. Such mistakes just mean that they are in the process of learning another language. If you speak slowly and clearly, it will help others to understand you. However, raising the volume of your voice will not accomplish anything.

Illiteracy. Offer help without making any judgement about the person's inability to read and write. Give clear, simple directions or draw a map showing landmarks. As with other disabilities, do not assume illiteracy means low intelligence. Be sensitive to feelings of embarrassment.

Fluency and articulation disorders. Try not to allow a stutter or a lisp to dominate your attention. You may end up missing what the person is saying and embarrassing him or her at the same time. Just as with a non-native speaker, remember that hesitant speech does not indicate slowness of thought. Be patient and relaxed, and listen carefully rather than attempting to speak for the other person.

Eye contact. Different cultures have different customs about eye contact. Some societies favor prolonged eye contact, while others find prolonged eye contact to be rude or threatening. If you notice that someone from another culture is either averting her eyes or holding them on yours for too long, keep in mind that she may be following different rules of etiquette. Generally, it is best to try to find a middle ground if you are unsure. Body language expert Julius Fast recommends that you break eye contact frequently as you talk or listen. Look down to the side and then back (Fast, 1994).

Hand gestures. As with eye contact, hand gestures are closely linked to culture. Some people use few hand gestures (such as the Japanese) and may seem stiff or standoffish to one who uses sweeping, expressive gestures (such as the Italians). Again, different rules of behavior may apply. Before making generalizations about any individual, try observing for a while.

Figure 12.7
Definitions of appropriate personal space vary from culture to culture.

Personal space. We may feel uncomfortable when strangers invade our "body space." However, how one measures the comfort zone depends upon one's culture. The normal social interaction distance can vary significantly among peoples (see Figure 12.7). Be aware if the person is instinctively inching closer to you or farther away, and try to accommodate his comfort zone. As with eye contact, when you are unsure, try a middle ground.

Conclusion

Communication allows the people of the world to share in the experience of being alive. It gives us the opportunity to grow and to evolve individually and collectively. "Communication is the back and forth of telling and listening and responding, so you know you are not alone" (Brandenberg, 1993, p. 3). We communicate in many ways, through our speech, our body language, and our written language. Our ability to communicate effectively may break down when we meet someone who has a physical impairment which creates a difficulty in speech or language, or when we are ignorant about the ways we can choose to send and receive information.

References

Brandenberg, A. (1993). *Communication.* New York: Greenwillow Books.

Beier, E. (1990). Body language. *Encyclopedia Americana* (International edition, vol. 4, pp. 131-32). Danbury, CT: Grolier.

Belinfante, A. (1997). Esperanto: What's that? [On-line.] Available: http://wwwtios.cs.utwente.nl/esperanto/baza_informilo/en.html)

Conley, C. (1997). *On top of the world: A life planner for every student.* Manuscript in publication.

Crystal, D. (1987). *The Cambridge encyclopedia of language.* Cambridge: Cambridge University Press.

Famighetti, R. (1996). *The world almanac and book of facts, 1997.* Mahwah, NJ: KIII Reference Corp, p. 646.

Fast, J. (1994). *Body language in the workplace.* New York: Penguin Books.

Gerbner, G. (1990). Communication. *Encyclopedia Americana* (International edition, vol. 7, pp. 423-24). Danbury, CT: Grolier.

Voice Foundation. (1995). *The voice foundation.* Philadelphia: The Voice Foundation.

Suggested Readings

Axtell, R. E. (1992). *Do's and taboos around the world* (2nd ed.) New York: Wiley.
This is an overview of intercultural communication and etiquette that helps you to avoid using offending or misleading gestures, body language, or phrases.

Berger, G. (1981). *Speech and language disorders.* New York: Franklin-Watts.
This is a readable explanation of the facts and fallacies of communication disorders.

Illich, I. & Sanders, B. (1988). *The alphabetization of the popular mind.* San Francisco: North Point Press.
This is an historical discussion of the development of written language and an exploration of how language alters our world view, our sense of self, and our sense of community.

Ong, W. J. (1982). *Orality and literacy.* New York: Routledge.
This book is a fascinating survey of primary oral cultures (those with no written language) and the societal effects of writing, print, and electronic technology.

Notes

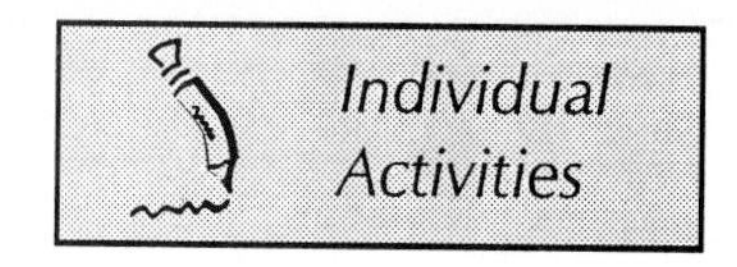

1. Imagine that you cannot read or are encountering an unknown language. How do you feel when confronted with the following message on a bottle?

Πθχλδλ Σ: Ωζδϖζλ! Ψζ λζβ Ψχδλη!

Reactions:

Now decipher the message using the key below:

θ ω Δ Ψ Φ ψ Σ α δ ϕ η φ κ λ ζ Ω ξ χ ϖ β μ Θ Π σ ϑ Ξ
a b c d e f g h i j k l m n o p q r s t u v w x y z

Reactions:

2. At lunch today, observe the body language of the people around you. Can you tell if someone is intrigued, or bored, or excited by a conversation? Do they touch each other, and if so, where and how often? Characterize their proximity to one another and thereby determine how well acquainted they are. Write your observations below:

1. a. Non-verbal visual communication serves a variety of functions. Facial signals are particularly versatile in responding to the speaker. As a group, complete the following list of facial expressions:

fear
happiness
anger

b. Certain body behaviors are used especially for ritual or official occasions. Complete the following list of ritual body language:

kneeling
bowing

2. Every language serves its speakers effectively. The Eskimos, for example, have numerous words for *snow,* but these would be totally unnecessary for the people of Egypt. As a group, come up with some words or concepts that are familiar to most Americans but that probably do not exist in another language like Chinese, Navaho, or Swahili.

Reflection Paper 12.1

Assume that you are a foreigner in a country with a different alphabet. Trying to decipher road signs would be quite a challenge. What are some other frustrating situations in which you may find yourself?

Reflection Paper 12.2

It has been suggested that learning another language teaches you how other people think. How is this so? Explain this concept based upon either your own foreign language study or the experience of someone you know who is bilingual.

Notes

Chapter 13 Behavior and Personality

Chapter Sections

- Introduction
- Personality Types
- Influencing Factors
- Childhood Patterns
- Inborn Traits
- Attention Deficit Hyperactivity Disorder
- Anxiety and Depression
- Suicide
- Punishment Versus Consequence
- Terminology
- Support Services
- Interactions
- Conclusion
- Individual and Group Activities

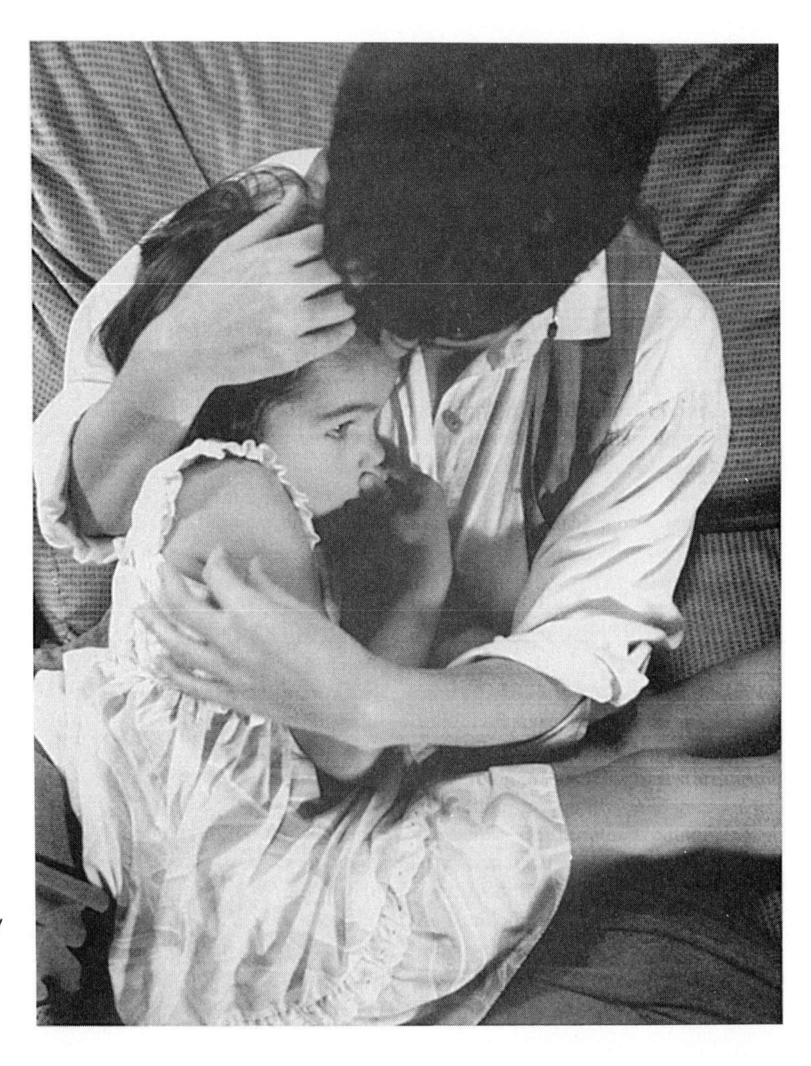

"Every individual on this planet is gifted with a truly unique personality."
—Jon Noring, scholar

Objectives

- Identify some factors which influence people's responses and behaviors.
- Explain what we mean by temperament.
- Explain how to more effectively deal with behavior and personality differences.
- Describe the identifying signs of hyperactivity.
- Identify the warning signs of suicide.

Assume you are a camp counselor and notice that Alex, one of your campers, is always isolated from the others. Troubled by this, you decide to appoint Alex as your helper. This attention aggravates him, sending him into a frenzy of kicking, biting, and yelling. Alex may have a behavior disorder. How would you deal with the situation?

Introduction

Among the categories of human diversity, people with behavior and personality disorders are perhaps the most misunderstood. Despite progress in many areas, the education, treatment, and integration of such persons into communities remain subject to debate among parents, community members, legislators, and educators. There is no denying that a small percentage of people who have behavior and personality disorders act in ways that are extremely offensive, even to the most tolerant among us. It is important to keep in mind, however, that this is true of a very small percentage of the population, and frequent interaction with such individuals is not likely. In this chapter, we will explore some behavior and personality differences among people which may affect their professional and social success, including anxiety, depression, mood swings, phobias, and suicidal tendencies. A familiarity with personality and behavior differences will equip you to identify problems and respond to them appropriately (see Figure 13.1).

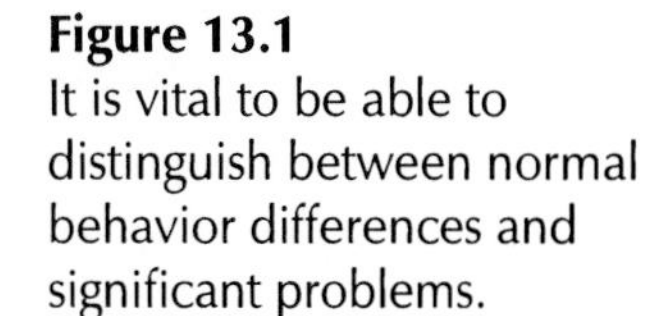

Figure 13.1
It is vital to be able to distinguish between normal behavior differences and significant problems.

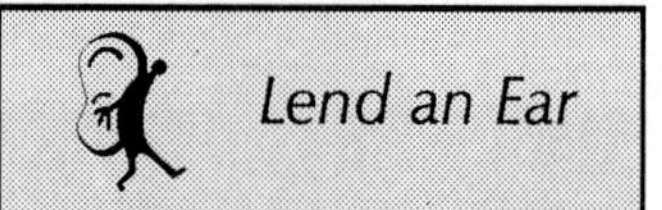

"From the moment of his birth the customs into which [an individual] is born shape his experience and behavior. By the time he can talk, he is the little creature of his culture."
—Ruth Fulton Benedict, anthropologist

Personality Types

As we observe individuals around us, we notice a great variety of personality traits among them. "Some people are very outgoing and fun-oriented, while others are more quiet and introspective; some people are highly analytical in decision making, while others use their feelings for deciding; some people feel more comfortable living a planned, orderly life, while others prefer to live spontaneously" (Noring, 1993, p. 1). Though no two people have exactly the same personality, behavior scientists have identified basic traits to describe people. For example, psychologists Katharine Briggs and Isabel Briggs-Myers have devised a famous and well-tested model of human behavior and decision making, categorizing people according to their tendencies in each of four behavioural characteristics (Fritz, 1996). The characteristics are as follows (adapted from Fritz, 1996):

1. *Introversion (I) and Extroversion (E)*
This trait is defined by the degree to which an individual understands his or her environment through careful consideration (I) versus understanding through externalizing and reacting (E).

2. *Sensing (S) and Intuition (N)*
This trait is defined by the way in which an individual perceives information. A sensor (S) determines information through careful observation. An intuitive (N) acquires information through relationships with people, patterns, and intuition.

3. *Thinking (T) and Feeling (F)*
This trait defines an individual's usual process of forming a judgment. The thinking end of the spectrum (T) indicates someone who is logical and deliberate in his or her decision making, relying on provable facts, whereas

the feeling end of the spectrum represents someone who relies mostly on how a decision will affect others when exercising judgment.

4. *Judgment (J) and Perception (P)*
People who fall in the judgment side of the spectrum are more comfortable living a life that is organized and manageable. They meet deadlines and don't always deal well with unexpected change. People who in the perception side of the spectrum are uncomfortable with too much structure in their lives. They do not like working under deadlines. They are often more adaptable than Js.

Everyone's personality can be described as a combination of these four categories. "No combination is better or worse, but brings different approaches and different qualities to work and decision making" (Fritz, 1996, p. 1).

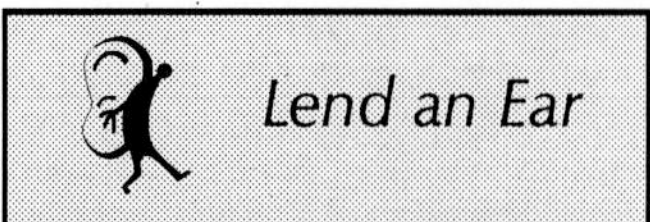

"Inappropriate behavior violates an unwritten social contract about how people should interact. We also find the aberrant behavior inherently fascinating."
—Belinda D. Karge, educator

Influencing Factors

We know that every person has his or her own personality traits and behavior styles. Such factors as birth order, age, gender, disabilities or health problems, family crises, sibling competitions, family values, socioeconomic status, and cultural and ethnic differences all affect reactions and responses. It is important not to overlook these differences when we consider the appropriateness of behavior in social settings.

Sometimes children grow up with good role models. Their parents, guardians, and other significant adults work hard at teaching acceptable behavior, but still the children behave in ways which are considered inappropriate as they grow up. Many people you meet, unfortunately, may not have had the benefit of adults in their lives who guided their behavior and served as good

Untrue Stereotypes of People with Behavior Differences:

- Are socially inept
- Are uncontrollable
- Are unstable
- Have emotional disorders as well
- Are all on medication
- Choose to be disruptive
- Are all "hyperactive"

role models (see Figure 13.2). Some people may come from homes where one or both parents were abusive, where older siblings were disruptive, and where socially unacceptable behaviors were the norm. People who exhibit disruptive behavior in public may be imitating the only model they have had.

Such disruptions as repeated outbursts, cursing, interfering with another's work, or physically or verbally attacking someone are some examples of behavior differences you might expect to encounter which would be disruptive enough to require action on your part. Others might be the failure of a person to interact at all, but to be noticeably withdrawn, to sleep at inopportune times, or to show signs of an eating disorder. A person might appear constantly nervous or anxious, or be fearful of speaking out in public. Certain life changes might trigger a temporary behavior problem in a person. The death or illness of a family member or friend, a divorce, a job transfer, a move, the birth of a child, or even a physical change such as new glasses may affect a person so that there is a noticeable difference in behavior temporarily. Some behaviors are clearly situational or temporary, but consistently unusual or inappropriate behavior needs help.

Figure 13.2
Not all children grow up with role models who work hard at modeling appropriate behavior.

Childhood Patterns

We will explore in depth some causes of behavioral differences later in this chapter. For now, suffice it to say that most patterns of behavior develop during an individual's formative years. The following disturbing statistics may help to illustrate some influences on today's young people. Every day in the United States over 1200 teenagers give birth, at least five teenagers commit suicide, over 1800 children are physically, emotionally, or sexually abused, over 3000 children run away from home, over 1500 children spend the night in a jail which was designed for incarcerated adults, and approximately 3000 children learn that their parents are going to be divorced (Eaton and Schwartz, 1996). You can see that problems such as academic underachievement, dropping out, drug and alcohol abuse, chronic physical and emotional problems, and, ultimately, suicide, may be the result of behavior and personality problems which have their roots in early childhood experiences.

As educated members of society, we are uniquely situated to observe these problems and to address them so that there is hope for positive change. It is vitally important, however, to keep in mind that we are not psychologists, physicians, or judges. What then, is our role in the life of a young person who shows signs of a behavior or personality problem? Our observations are valuable, and we can play a crucial part in changing social behavior and in giving children a new sense of themselves by the way we interact with them. However, it is not our role to treat illnesses, mental or physical, nor to diagnose any child. In cases where treatment is needed, our role is to refer the individual for appropriate help to a physician or mental health professional.

Figure 13.3
Every person has an inborn temperament, affecting one's outlook and behavior.

Inborn Traits

In order to distinguish between normal behavior differences and significant problems, we first need an understanding of how human beings all differ. Every individual has a natural, inborn style of behavior, which we can call his or her temperament (see Figure 13.3). This is not associated with motivation, but rather is a description of *how* an individual behaves in a certain circumstance. We can influence temperament by our interactions and by control of the environment, but we cannot change or cause temperamental characteristics (Turecki & Tonner, 1985).

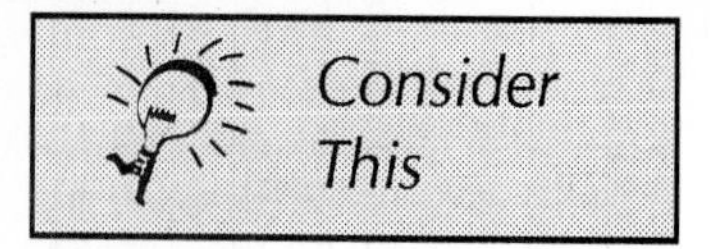

Assume that you have just been told by a psychologist that your child has a behavior disorder. How would you react to the diagnosis? How would you feel? What would you tell others?

Dr. Stanley Turecki (1985), a noted child psychiatrist, has described nine traits which we can use to identify and describe the way a child interacts with his or her environment:

1) Activity level. How active is the child generally?

2) Distractibility. Can the child pay attention? How easily is he or she distracted?

3) Persistence. Can the child stay with something he or she likes?

4) Adaptability. How does the child deal with change, or with transition from task to task?

5) Approach/withdrawal. How does the child initially react to newness?

6) Intensity. How loud is the child usually, whether happy or unhappy?

7) Regularity. How predictable is the child in physical habits?

8) Sensory threshold. How does the child react to noise, lights, smells, or temperature?

9) Mood. Is the child predominately negative or positive?

(Turecki & Tonner, p.14)

Clearly, children with lower activity levels, less distractibility, more adaptability, less intensity, and more positive moods interact with others more easily. However, those with the opposite, "difficult" attributes are not necessarily abnormal, according to Dr. Turecki. "Abnormality," he explains, "implies the presence of a clear diagnosable disorder. Human beings are all different, and a great variety of characteristics and behaviors falls well into the range of normality" (Turecki & Tonner, 1985, pp. 15-16).

Attention Deficit Hyperactivity Disorder

Most children who have diagnosed behavior or personality differences will have been assessed for placement in special education, or have prescribed medications, or both. The best known developmental disorder of the eighties and nineties is Attention Deficit Disorder, or Attention Deficit Hyperactivity Disorder (ADHD). This disorder can affect a child's school performance, self-esteem, and ability to behave in appropriate ways in a social or academic setting. Some researchers believe that ADHD is simply a normal human variation, the way that height, intelligence, or athletic ability are. Accordingly, in a typical classroom there will be students who can concentrate all day long, and some who are extremely easily distracted from their tasks (Garber, Garber, & Spizman, 1995).

ADHD differs from a learning disorder in that ADHD is usually diagnosed on the basis of the individual's ability to cope with everyday life rather than by a standardized test. It is characterized by inattention, impulsivity, and hyperactivity (Weaver, 1995). Typical behaviors of a person with ADHD include fidgeting, blurting out answers, difficulty waiting for his or her turn, talking excessively, being easily distracted from a task, interrupting, and difficulty sitting quietly.

According to Dr. Stephen Garber (1995), problems of the child with ADHD may be exacerbated if he or she is required to participate in the wrong kind of experience. Garber says that more structured settings are best for the child with ADHD, and that routines and consistent consequences, both negative and positive, are

First Person

I haven't always known about ADD, and when I first learned about it I thought it was an excuse for poor behavior. My younger brother, 13, and my father, 45, both have it. It's not easy for me to accept; in fact, we often end up in a fight about it sooner or later. People with ADD can be very impulsive and forgetful, and it's frustrating to deal with them on a daily basis. You must constantly check up on them. You also need to repeat a lot of things, so it makes you feel as though they are not listening to you, and then you get angry. When my brother finally sits down to do his homework we have to leave him alone. We aren't allowed to ask him questions or do anything that might distract him.

I have a lot of the same problems that my dad and brother have, like lack of concentration, misplacing things, and forgetting what I was in the middle of doing, but I feel that I haven't let them get in my way. My parents suggested that I get tested for ADD, but I honestly feel, in my own insecurity, that I'd rather not know. I continue to try to learn more about ADD to be able to deal with it better.

—K. C., New Brunswick

best. Whenever possible, distractions should be minimized, and tasks broken into smaller parts (Garber, Garber, & Spizman, 1995). The role of parents, teachers, and other caregivers is to help each child find the optimal way of learning, both at school and at home, while establishing and maintaining an orderly environment. Be ready to refer any children whose needs are clearly not being met for expert evaluation and help.

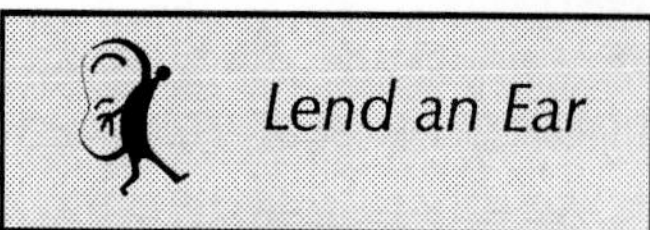

"My son still does some bizarre things, but now that I know he has a behavior disorder, I don't go into a panic anymore. I simply address the problem behavior with him, and my family lives from crisis to crisis."
—Anonymous.

Anxiety and Depression

When a person often seems overly worried in a nonspecific way, or is in constant need of reassurance, there is a possibility that he or she is experiencing anxiety. Anxiety affects an individual's ability to succeed in life because he or she cannot concentrate. It inhibits the individual's ability to enter into new experiences, and it negatively affects self-esteem (Doft, 1992). Creating situations where the individual can succeed and gradually grow in confidence may help. Small steps in mastery can lead to a willingness to try new things. However, if you observe that the person's anxiety does not lessen, or that it interferes with his or her social life, you should suggest a professional evaluation (Doft, 1992).

Another common disorder is depression. Depression does not affect only adults. Even the youngest child may suffer from depression. An individual who is listless, seems uncharacteristically sad, has trouble concentrating, or is acting out in ways that are not typical of him or her, may have depression. Individuals may be depressed if they have a persistently sad facial expression and demonstrate many of the following symptoms nearly every day for a minimum of two weeks (Rosenberg,1995):

- poor appetite
- sleep changes

• motor restlessness
• lack of activity
• loss of usual interests
• signs of apathy

Sometimes depression is also masked as anger, and attention should be paid to the person who is exhibiting any of these symptoms. If you notice persistent symptoms which you cannot identify as situational, (for example, a family crisis) then you should take them seriously, and seek help for person (Doft, 1992).

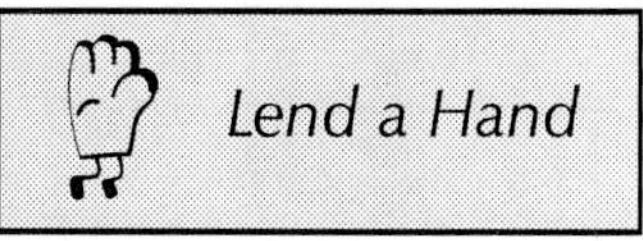

Disruptive behavior is often an expression of frustration. Honor personality and behavior differences by involving people in projects that allow many alternative modes of expression, such as art, language, and music (Perkins, 1992).

Suicide

Depression is sometimes associated with suicide. Even the youngest child may talk of suicide, or engage in very risky, dangerous behavior such as jumping off a building. You should know some warning signs which may signal distress so severe that it could lead to suicide. Here are some of the most common warning signs (adapted from Kolehmainen & Handwerk, 1986, p.15-17):

1) The person has previously attempted suicide.

2) The person talks about death or suicide.

3) There is a noticeable change in the personality or mood of the individual, for which there is no obvious explanation.

4) There are changes in sleep and eating patterns, which may be noticeable at school.

5) The person in distress may withdraw from friends and his or her usual activities.

6) The person who is suicidal may take unusual risks,

What are the warning signs of suicide?

such as driving recklessly, which show a disregard for life.

7) There may be uncharacteristic drug or alcohol abuse, or abuse of prescription drugs.

8) The person may make final arrangements, write farewell letters, give away possessions, or attempt to conclude unfinished business.

If you have observed some of the above warning signs in a person, one of the best approaches is to ask questions. Kolemainen & Handwerk (1986) explain that questions allow you to get information you need to make an assessment of the situation and show your interest at the same time. Ask the person such questions as why he or she is so unhappy, whether anyone else knows about this unhappiness, what would make things better, and what you can do to help right now. Ask questions and listen rather than trying to offer answers and solutions (see Figure 13.4). Involve professionals by referring the person to a counselor or crisis intervention agency immediately, and follow up to be sure someone is helping (Kolemainen & Handwerk, 1986).

Figure 13.4
Asking questions is a vital step in helping someone who is considering suicide.

Punishment Versus Consequence

When people violate rules of behavior or act inappropriately, whether because of a personality or behavior difference or because of a particular situation, you may feel compelled to respond in some way. That response may result in a consequence or in a punishment. Curwin & Mendler (1990) have explained that consequences teach responsibility, while punishments teach obedience through fear. Punishments include measures such as threats, lectures, loss of privileges, scolding, or humiliation. Punishments are not directly related to the

What is the difference between a punishment and a consequence?

rule which has been violated. Consequences follow naturally and logically from the person's actions. They are directly related to the rule which has been broken, and that connection is obvious to the person (Curwin & Mendler, 1990). The following chart illustrates the difference:

Punishment	Consequence
Attacks person's dignity	Enhances person's dignity
Is externally oriented	Is internally oriented
Focuses on the past	Focuses on the future
Gives short-term results	Gives long-term results
Is not related to the rule	Is related to the rule

(Adapted from Curwin & Mendler, 1990, pp.56-57)

Such things as berating, isolating, and threatening the loss of privileges are punishments. Consequences teach the person a lesson he or she can internalize, without a loss of self-esteem. For example, a missed appointment might mean making it up during what is usually free time. That consequence would encourage the person to be more responsible, without labeling his or her behavior negatively.

Curwin & Mendler (1990) have suggested the use of three kinds of consequences: predicting, choosing, and planning. *Predicting* is simply asking the person what will happen in the future if he or she repeats the inappropriate behavior. *Choosing* is giving the person a choice from several alternative behaviors. *Planning* is asking the person for a solution to the problem (Curwin & Mendler, 1990, p.57-59). This approach is an effective means of encouraging appropriate behavior because it involves the person in the solution to the problem. Of course, it is necessary to tailor this response according to particulars of the situation.

Personal Checklist

No one is perfect. If your attempts to redirect misbehavior sometimes fail, you may find this checklist useful:

___ I made sure that I did not make someone suffer.

___ My tone of voice was not condescending.

___ The person felt liked and accepted by me.

___ I encouraged the person to voice his or her opinion.

___ I respected the person.

___ I recognized the purpose for the misbehavior.

___ The person did not feel overpowered by me.

___ I avoided talking too much.

___ The consequence was logical.

___ I presented an alternate way for the person to feel special.

(adapted from Kvols-Riedler, 1979, p. 244)

Consequence-oriented responses might not accomplish your goal in every circumstance. Sometimes people must immediately follow your instruction, with no explanation at all. For example, when safety is involved, such as during a fire, a sudden emergency, or an assault upon one person by another, there may be no time for discussion.

Terminology

Following are some common terms relating to behavior and personality differences that you may encounter and should be familiar with:

Autism. This is an incapacitating disability affecting nonverbal and verbal communication and two-way social interaction. It affects one out of every 2,500 children. Typical characteristics include repetitive movements, abnormal responses to sensory stimulation, avoidance of eye contact, and insistence that routines remain unchanged.

Bipolar depression. This is characterized by high-to-low mood swings, impulsivity, and high distractibility.

Disruptivity. A person who is disruptive may start fights, demonstrate self-abusive behavior, try to draw attention to himself or herself by making noises, or generally generate a loss of order in the environment.

Manic Depressive Disorder. Manic depressive symptoms include episodes of extreme sadness, hopelessness, chronic fatigue, and irritability.

Mood swings. This refers to patterns of significant change in values, attitudes, compliance, friends, attire, and/or priorities.

Phobic disorders. A person with a phobic disorder has an unreasonable, recurring fear of a specific object, activity, or situation. Most people have some mild phobias, such as fear of heights, certain animals, flying, or public speaking. When a fear drastically interferes with everyday life, a phobia is present.

Schizophrenia. This should not be confused with multiple personality disorders. Schizophrenia involves a disturbance in thinking patterns that causes a person to act and speak strangely. People with schizophrenia may experience hallucinations and delusions.

Temperament. This refers to character traits with which we are born. Sometimes we refer to temperament as disposition. It is the basis of our personality.

Tourette's Syndrome. This is an inherited, neurological disorder involving involuntary, rapid, nonrhythmic motor movements or vocalizations.

Issues Persons with Behavior Disorders May Face:

- Others considering them "difficult" to get along with
- Ridicule from others
- Classroom problems
- Finding acceptance
- Finding support
- Finding good role models
- Difficulty with relationships
- Labeling
- Finding the "right" type of help
- Getting a job that allows for behavior differences
- Other people's perceptions

Support Services

Behavior therapy is a valuable, relatively inexpensive treatment for most of the topics covered in this chapter. Psychiatric treatments include medicine, psychotherapy, psychoanalysis, and group and individual counseling. Psychiatrists, trained in both physical and psychological illness, can diagnose the symptoms and prescribe the appropriate treatment. Psychologists, trained in a variety of non-medical intervention strategies, may provide excellent therapy options. Other professionals, such as behavior specialists, art therapists, and music therapists, also offer excellent therapeutic interventions. Support groups can be a critical part of coping with behavior and personality differences. Groups are available for the person with the disability as well as for family members and friends.

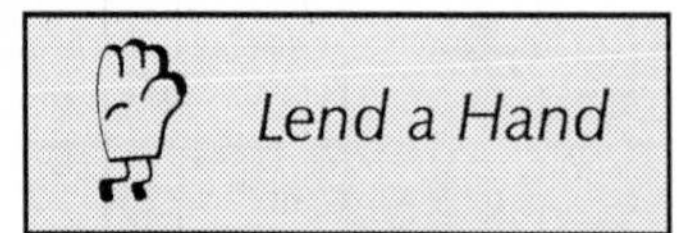

Provide opportunities for the person to be helpful, to cooperate, to participate, and to do what he or she can do to make the situation more enjoyable (Kvols-Riedler, 1979).

Interactions

Keep your sense of humor. Humor reduces stress and tension (Curwin & Mendler, 1990).

Examine your attitudes about control and discipline. Do you believe that people respect only what they fear? Do you think you must be tolerant of rude or disruptive behavior? (Kvols-Riedler, 1979).

Avoid power struggles. Respond with action and limit your verbal responses (Curwin & Mendler, 1990). This will help you avoid turning disruptive situations into a battle of wits.

Establish and maintain a relationship of equality and mutual respect (Kvols-Riedler, 1979). Such a relationship will create a safe environment for all people.

Establish clear, specific rules. Stick to these, but limit them to the necessary minimum (Curwin & Mendler, 1990). Know when it's appropriate to bend the rules (Weaver, 1995).

Have more than one perspective. Ask yourself how the person sees the situation. Consider possible causes of his or her behavior (Kvols-Riedler, 1979).

Give positive reinforcement at every opportunity (Curwin & Mendler, 1990). Never miss an opportunity to reward good behavior.

Gain insight into the person's mistaken goal. Then help the person identify his or her mistaken goal in a nonaccusing way. Watch the person's facial expressions and body language to gauge whether your corrections are discouraging or frightening, rather than informing (Kvols-Riedler, 1979).

Conclusion

As Dr. Richard Curwin and Dr. Allen Mendler (1990) have written, it is of utmost importance to remember that concentrating on disabilities and weaknesses may worsen a person's poor self-esteem and contribute to greater social problems and professional failures. Attaching negative labels to those who are different and attempting to remediate them into being like everyone else can be extremely damaging, according to Curwin & Mendler. It is important to give positive messages about what people can do rather than what they cannot do. Our focus must be on their strengths, not their weaknesses (Curwin & Mendler, 1990).

References

Curwin, R. L. & Mendler, A. N. (1990). *Am I in trouble? Using discipline to teach young children responsibility.* Santa Cruz: Network Publications.

Doft, N. (1992). *When your child needs help: A parent's guide to therapy for children.* New York: Crown Trade Paperbacks.

Eaton, L, & Schwartz, S. (1996). *Exceptional people study guide.* Gainesville, FL: Department of Independent Study by Correspondence.

Fritz, J. (Sept. 1996). Myers-Briggs personality index. [On-line.] Available: http://wwwos2.cs.unb.ca/profs/fritz/cs3503/person35.htm

Garber, S. W., Garber, M. D., & Spizman, R. F. (1995). *Is your child hyperactive? Inattentive? Impulsive? Distractible?* New York: Villard Books.

Kolehmainen, J. & Handwerk, S. (1986). *Teen suicide.* Minneapolis: Lerner Publications Company.

Kvols-Riedler, B. (1979). *Redirecting children's misbehavior: A guide for cooperation between children and adults.* Gainesville, FL: National Parenting Instructor Network.

Perkins, D. (1992). *Smart schools: From training memories to educating minds.* New York: The Free Press.

Noring, J. (1993). Personality type summary. [On-line.] Available: http://www.pendulum.org/misc/mb.htm

Rosenberg, B. A. (1995). *Depression in children.* Philadelphia: Cornerstone Psychiatry Associates.

Turecki, S. & Tonner, L. (1985). *The difficult child.* New York: Bantam Books.

Weaver, C. (1995). *Success at last! Helping students with attention deficit (hyperactivity) disorders achieve their potential.* Portsmouth, NH: Heinemann.

Suggested Readings

Briggs-Myers, I. & McCaulley, M. (1985). *Manual: A guide to the development and use of the Myers Briggs type indicator.* Gainesville, FL: Consulting Psychologists Press.

This manual is an essential reference for understanding and using personality type indicators.

Duke, P. & Hochman, G. (1992). *A brilliant madness: Living with manic depressive illness.* New York: Bantom Books.

In this book, Patty Duke joins with medical reporter Gloria Hockman to explain the powerful, paradoxical and destructive manic depressive illness.

Elliott, M. & Meltsner, S. (1991). *The perfectionist predicament: How to stop driving yourself and other people crazy.* New York: W. Morrow.

Provided in this book are helpful ideas for working with a perfectionist.

Gehret, J. (1991). *Eagle eyes: A child's view of attention deficit disorder.* Fairport, NY: Verbal Images Press.

This book presents a picture of living with ADD from the point of view of a child who lives with the disorder.

Individual Activities

1. Positive reinforcement is often used to control or modify the behavior of people with emotional disturbances. List 10 reinforcements which would be appropriate to use with children, and 10 for adults.

2. Find out how others perceive emotional or behavioral disturbance by asking five of your friends for a definition. Write their definitions. Rate each definition's accuracy (A, B, C, D, or F).

1. There has been conversation regarding the opening of a center for people with behavior and personality disturbances in your neighborhood. Form groups of 4-5 people and discuss the pros and cons of the center and its location. Then present your group's ideas to the entire class.

Reactions:

2. With your same group of 4-5 people, develop a list of terms which have been used to label people with personality or behavioral disorders. Identify those terms which are positive and those which are negative.

Reactions:

Reflection Paper 13.1

If a child of yours or a good friend began to display characteristics of behavior disorders, suggest ways that you would effectively deal with them.

Reflection Paper 13.2

What are some warning signs of suicide, and how would you help a person who is considering suicide?

Notes

Chapter 14 Sensory Differences

Chapter Sections

- Introduction
- The Culture of Blindness and Deafness
- The World of Sound
- Hearing Impairment
- Terms Related to Hearing
- Support Systems for Hearing Loss
- Interactions
- The World of Vision
- Visual Impairments
- Terms Related to Vision
- Support Services for Vision Loss
- Interactions
- Conclusion
- Individual and Group Activities

"The responsibility for tolerance lies in those who have the wider vision."
—George Eliot, author

Objectives

- Identify factors that can disrupt the hearing process.
- Explain how hearing loss is classified.
- Distinguish "deaf" from "hard of hearing."
- Identify three causes of visual impairment.
- Explain how a visual impairment is classified.
- Define legal blindness.

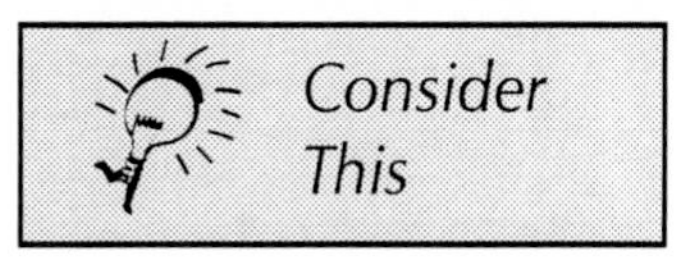

If you had to choose being blind from birth or acquiring blindness later in life, what would you decide, and why?

Introduction

We use our senses—hearing, sight, smell, touch, and taste—to gather the information which we require to function safely and effectively in the world (see Figure 14.1). Author and lecturer Helen Keller rejoiced in sensory experience. She wrote at length about life's abundant aromas, tastes, touches, and feelings. The fact that she could neither see nor hear did not diminish her zest for the sensory world. She was able to enjoy music by placing her hands on a radio, to read literature in Braille, to communicate with her friends through sign language, and to write down her wisdom and experiences. As Helen Keller demonstrated by expressing her ideas and emotions, people with hearing or vision impairments do not think differently than other people. Nor do they necessarily lead impoverished lives. The purpose of this chapter is to examine the broad range of hearing and vision impairments and suggest methods for fuller, more enjoyable, and more effective communication with those individuals.

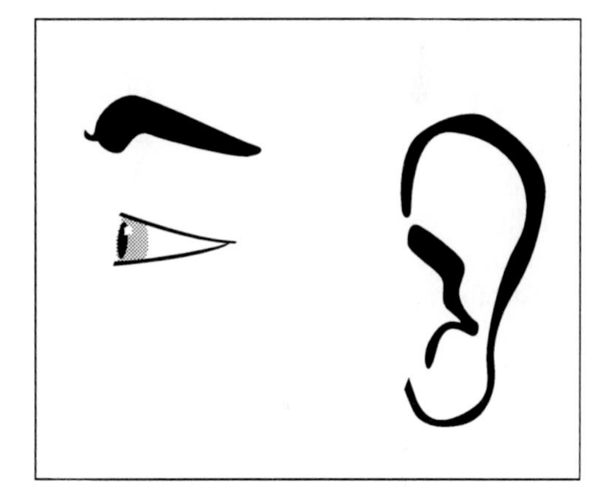

Figure 14.1
Though our senses help us to gather vital information, everyone has different ranges of hearing and vision.

The Culture of Blindness and Deafness

Hearing is the sense we use primarily in the development of language and speech. Since speech is the means by which we communicate with others, it is a fundamental factor in all social interactions. Hearing facilitates learning, and through hearing and speech we pass on our cultural values and our heritage. Vision, too, is important for assimilating our culture. We read printed words in books, watch the moving images in films, and study the intricate details in paintings. As with any culture, persons from deaf and blind cultures celebrate their ancestors, heroes, victims, survivors, and trailblazers.

The culture of deafness is quite rich. From the ancient Greek historian Herodotus to the French novelist Guy de Maupassant, people have written eloquently about their own deafness or the deafness of friends and loved ones (Ackerman, 1990). Brian Grant's anthology *The Quiet Ear* compiles writings about deafness that span many different eras and cultures, and the play *Children of a Lesser God,* by Mark Medoff, has been made into a powerful movie. The German composer Ludwig van Beethoven, who became totally deaf at age 46, wrote his greatest music during his later years (see Figure 14.2).

Figure 14.2
Beethoven continued composing famous musical works after he lost his hearing.

Just as with deafness, blindness has not hindered the productivity of some of our great cultural figures. The Argentine poet and story writer Jorge Luis Borges, many of whose works have been translated into English, was blind. Joseph Pulitzer, the prominent journalist, publisher, and congressman, went blind at the age of 40 but continued his various activities during the remaining 24 years of his life. James Thurber, the well-known magazine writer, dramatist, and cartoonist, lost the sight of one eye in a boyhood accident and the sight of the other as an adult. And the Greek poet Homer, author of the famous epics *The Iliad* and *The Odyssey*, was blind.

Let's examine the senses of hearing and vision in a little more depth, each in turn.

The World of Sound

How do you extend your range of hearing every day?

What exactly does *hearing* mean? To a porpoise, hearing is a kind of sonar, like a bat's, that brings back three-dimensional images more like sights than sounds. A porpoise can "hear" all sorts of details about a shark—its size, texture, motion, direction, and distance. If you go to a rock concert or play a stereo loudly, you may *feel* the pulsing rhythm vibrating in your chest cavity. What hearing entails often depends upon the hearer and the context. Our range of hearing also depends upon which tools we use to extend it (see Figure 14.3). A stethoscope allows us to hear someone's heart beating. Loudspeakers make it possible to hear an orator in a large auditorium. With a telephone or appropriate computer software, we can hear someone in another part of the world. A radio telescope allows us to hear the distant echoes of outer space. And a hearing aid amplifies the volume of the sounds around us.

Of all the senses, hearing offers perhaps the greatest potential for information because of its flexibility. While taste, touch, vision, and to a lesser extent smell require our proximity to the source, hearing potentially offers information about objects and events when all other senses are useless. For instance, we may hear a siren in the dark outside from our bedroom and be able to tell that there is an emergency or a fire nearby. We may be able to judge the distance and direction, and even determine if the siren is from a fire truck or police car, fairly accurately by using our sense of hearing.

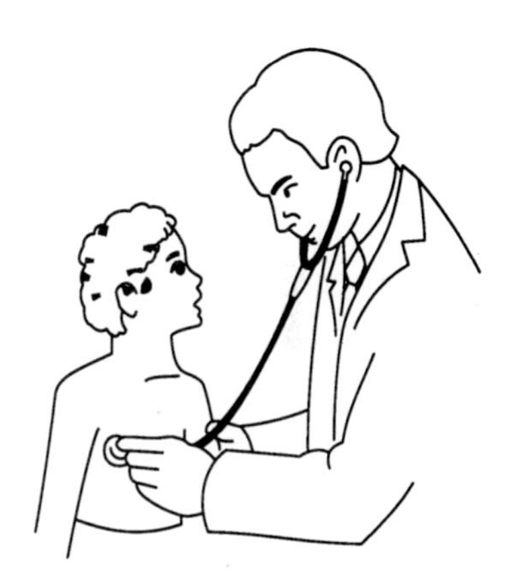

Figure 14.3
Can you hear someone's heart beating? Various tools extend our range of hearing.

Nobody can escape the world of sound. As sensory expert Diane Ackerman observed, even if we don't hear the outside world, we hear the throbbings and buzzings and whooshings of our own bodies. Many who are legally deaf can hear gunfire, low-flying airplanes, jackhammers, motorcycles, thunder, and other loud noises. Hearing disabilities don't protect us from ear distress, either, since we use our ears for more than just hearing. Our ears help us keep balance and equilibrium and tell the brain how our head moves (Ackerman, 1990).

Hearing Impairment

Our ability to hear is the result of a complex sequence of events. The outer ear collects sound waves and channels them through the auditory canal. The eardrum conducts sound waves through three tiny bones in the middle ear. The third bone is connected to the inner ear, which houses the cochlea. Here highly specialized cells translate vibrations into nerve impulses that are sent directly to the brain (see Figure 14.4).

Estimates of hearing loss in the United States go as high as 28 million people, or 11 percent of the total population (Toufexis, 1991). Our range of hearing can be affected by a variety of circumstances. Factors present before, during, or after birth can disrupt the hearing process. Childhood ear infections, loud concerts, gunshots, fireworks, and loud noises at work all have the potential to reduce one's hearing acuity. Even the natural process of aging gradually reduces our ability to hear high frequencies. According to researchers at Idaho State University (1996), ninety percent of young children's knowledge is attributed to incidental reception of conversations around them. Thus, learning is potentially hindered even with the slightest hearing loss.

Figure 14.4
Sound is processed through several intricate mechanisms in the inner ear.

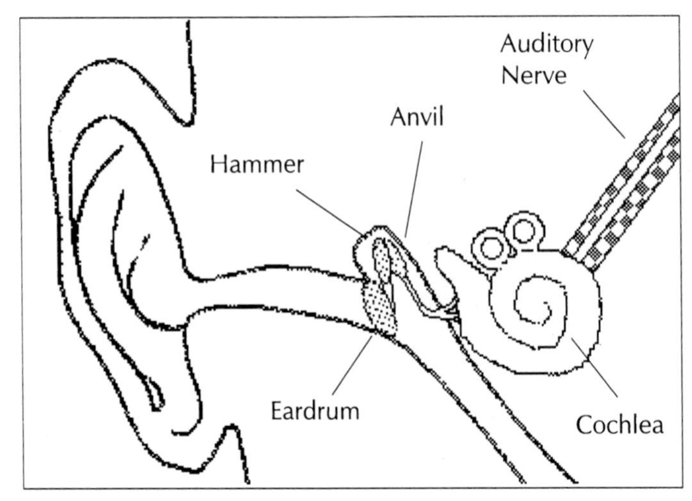

There is no single phenomenon of "deafness," but rather a wide spectrum of hearing loss, from mild impairment to total deafness. We use the term *hearing impaired* to refer to all individuals who have a hearing impairment, regardless of its severity. A hearing loss is reported in decibels (dB) and is generally categorized as:

- mild (27-40 dB)
- moderate (41-55 dB)
- moderately severe (56-70 dB)
- severe (71-90 dB)
- profound (91+ dB)

A person with mild hearing loss may experience difficulty with faint or distant speech. A moderate loss makes speech beyond five feet hard to understand. A person with severe loss is unlikely to hear a loud voice if it is more than one or two feet away, though he or she may be able to distinguish between different environmental sounds. Finally, a person with profound loss may be able to hear only very loud environmental sounds.

People with hearing impairments communicate through speech, lip or speech reading, or sign language. Depending upon the severity of hearing loss, one may use hearing aids to amplify sounds or have an electronic cochlear implant which aids the individual by stimulating nerve endings when sound is perceived. Obviously, hearing loss increases the challenges a person faces in daily living, but it does not necessarily prevent one from enjoying life or attaining his or her goals.

Untrue Stereotypes of Persons with Hearing Impairments

- People with profound deafness cannot learn to speak
- People with hearing impairments are not as challenged as those with visual impairments
- All people with hearing impairments can read lips
- The inability to hear is a sure sign of aging and/or senility
- People who use sign language simply gesture to one another in a very limited form of communication
- People with hearing impairments are usually also mentally retarded

Terms Related to Hearing

Following are some common terms relating to hearing differences with which you should be familiar.

Bilateral loss. This is a hearing impairment in both ears.

Conductive loss. This impairment is located in the outer or middle ear. With proper amplification, sounds may be heard without distortion.

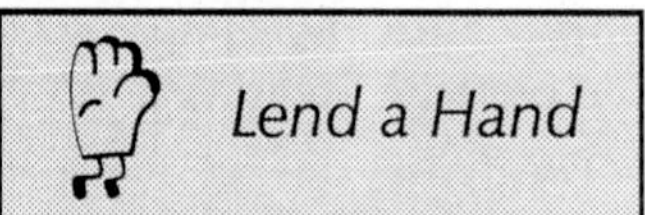

Learn to say "I would like to communicate with you" in sign language.

Issues that Persons with Hearing Impairments May Face

- Comprehension and production of spoken language
- Lower academic achievement
- Social isolation
- Separate or regular education
- Finding good auditory training
- Restrictions on the choice of job due to perceived limitations
- Adaptable equipment
- Finding support
- Finding role models and mentors
- Dealing with other people's discomfort
- Feelings regarding the need for assistance
- Independent living

Deaf. This is a hearing loss so severe that it generally is not helped by amplification. An individual with total deafness must rely upon vision as the primary means for developing communication.

Hard of hearing. This refers to a hearing loss that does not prohibit the development of speech and language skills, with or without amplification.

Postlingual hearing loss. A hearing impairment that occurs after speech and language skills have been acquired is called "postlingual." An individual with postlingual loss may have no difficulty with speech and language.

Prelingual hearing impairment. A hearing impairment that occurs either at birth or before speech and language skills have been acquired is called "prelingual." An individual with prelingual loss may have difficulty mastering language and speech.

S**ensorineural loss.** This hearing impairment is located in the inner ear or along the auditory nerve to the brain. Even with amplification, sensitivity to sounds is reduced and/or distorted.

Tinnitus. This refers to a range of noises in the ear (typically a ringing or hissing) that can disrupt hearing.

Unilateral loss. This is an impairment in only one ear.

Support Services for Hearing Loss

Telecommunication devices (TDDs), resembling small typewriters, can be hooked up to telephones to allow people who are hearing impaired to carry on a conversation. Captioned films and close-captioned television, now offered routinely with new television sets, offer subtitled dialogue. Signaling devices, using vibrations or flickering lights, can alert people with hearing impairments when the doorbell rings, the alarm clock goes off, or the baby cries. Fire alarms in buildings often have strobe lights to alert those who may be unable to hear standard bells or sirens. Telephone companies also offer "Relay Services" which allow an individual with a hearing impairment to engage in communication with a hearing individual in a manner equivalent to those individuals who are able to use standard voice telephone services. Relay Services utilize employees who use TDDs the same way an interpreter would.

Sign language classes are frequently included in community education programs offered by high schools, community colleges, and universities. Police and fire departments often train in basic sign language to be better prepared to communicate in any situation. Sign language interpreter services are typically offered in places of worship and at public events. Interpreters are also on-call 24 hours a day in many communities to assist in emergencies such as accidents and arrests, and 911 centers are required to have TDDs.

Interactions

Learn sign language. If you are skilled in some basic signs, you'll be able to communicate more comfortably and fully.

A light touch on the arm is appropriate to get someone's attention. To avoid possible offense, do not touch other parts of the body and do not wave your hand in an exaggerated fashion.

Make sure the individual with a hearing impairment can see your face clearly. Avoid turning away, moving around excessively, or obstructing your mouth.

First Person

My child, Juanita, was born with normal hearing. Today, she cannot hear a single word, but she lives an independent and fulfilled life. At age seven she developed spinal meningitis, which resulted in a severe hearing loss in both ears. Her teachers, mother, and I all worked closely with her to retain the speech that she had acquired prior to the loss.

Juanita spent the first half of her elementary school years in a special program for children with hearing impairments. Then she entered a regular classroom. As an adolescent, she graduated with honors from the neighborhood high school. She decided to continue her education at the university level, earning an undergraduate degree in international relations and then completing law school.

Now Juanita is an attorney. She actively volunteers for local charities and spends her leisure time doing photography, collecting stamps, and rollerblading.
—Rodriguez M., Chula Vista, CA

Speak in moderate tones. Raising your voice or exaggerating your speech may only distort your facial expressions and make you more difficult to understand.

If you are not understood, rephrase rather than repeat what you said. Rephrasing your thoughts may make your point more clear, and the person will have a second chance to understand you.

If you are in a group, speak one at a time. Too many people talking at once can be confusing to anyone, with a hearing impairment or not.

If you are communicating through an interpreter, speak directly to the person with the hearing impairment. Keep in mind with whom you are talking. The interpreter is just there to interpret.

The World of Vision

Figure 14.5
The eye translates photons of light into visual images.

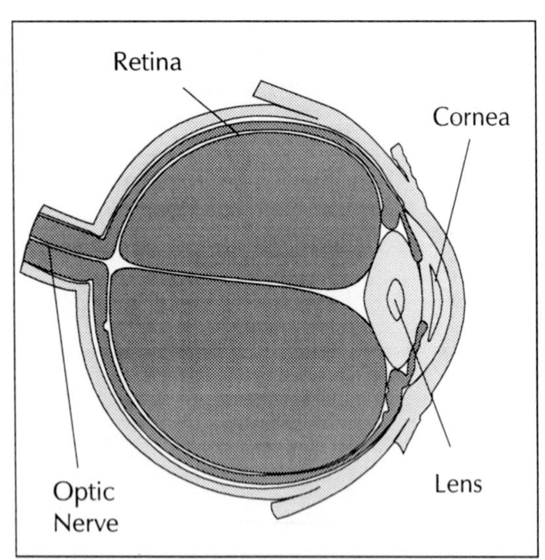

In the absence of light, everyone is blind. Photons of light bounce randomly off the objects around us, but it is the millions of rods and cones in the human retina that capture the light and send it through the optic nerve to the brain (see Figure 14.5). Then the brain translates the photons into meaningful vision. Through the visual process, we observe the world around us and assimilate knowledge. We rely upon our eyes to direct us through our environment, to inform us through the written word, and to give us pleasure. But eyesight isn't the only means by which we can perceive the world.

What do you see when you walk into a room? An old friend? A boring co-worker? A beautiful landscape painting? An expensive vase? Much of what we "see" is wholly subjective. Our eyes do not technically "see" an old friend, but rather the figure of a person. We "see" the relationship in our minds. Whether the painting is beautiful or not, and whether the co-worker is boring or not, are also opinions we form in our minds (Chopra, 1991). What we see with our eyes is certainly important, but visual information is clearly only one part of observation. A person with no eyesight might touch your face and "recognize" that you are her friend or think that you resemble someone else. Blindness and visual impairment affect *how* information is obtained, but not necessarily *what* information is obtained.

Though blindness conjures up images of total darkness in the minds of the general public, only a small number of people are totally blind. Even people who have been totally blind since birth are greatly affected by light, Ackerman noted, because light influences us in many subtle ways. "It affects our moods, it rallies our hormones, it triggers our circadian rhythms [biological cycles recurring at 24-hour intervals]" (1990, p. 249).

How do you extend your field of vision every day?

Visual Impairments

We extend our field of vision in all sorts of ways, using glasses, contact lenses, magnifying glasses, telescopes, cameras, binoculars, microscopes, computers, X-rays, and magnetic resonance imagers, just to name a few (see Figure 14.6). Some people must augment their vision due to a physical problem with their eyes. At least twenty percent of the population has some visual problems (Reynolds and Birch, 1982), but most of these cases can be corrected to the extent that the problem is not serious. It is estimated that approximately 1,440,000 individuals of all ages have visual impairments that are significant enough to limit their activities (LaPlante, 1991). The figure of one-tenth of one percent of the population is frequently cited for the prevalence of those people legally blind.

Figure 14.6
Every day, people extend their field of vision with various technological devices.

Untrue Stereotypes of Persons with Visual Impairments

- All people who are blind have superior musical talents
- People with visual impairments automatically develop heightened senses of smell, touch, hearing, and taste
- People with visual impairments are able to detect obstacles with a "sixth sense"
- People who are legally blind have no functional vision at all
- People with visual impairments cannot travel without assistance

The term *visually impaired* describes a wide range of people with partial or complete loss of sight. Visual impairment may be either present at birth or acquired later through injury to the eye or brain. Visual disability may be due to:

- refractive problems (farsightedness, nearsightedness, blurred vision, cataracts)
- muscle disorders (uncontrolled rapid eye movements, crossed eyes)
- receptive problems (damage to the retina and optic nerve).

Visual impairment is classified into three general categories: profound, severe, and moderate. With profound visual disability, one's performance of the most basic visual tasks may be very difficult. With severe visual disability, extra time and energy are needed to perform visual tasks. With moderate visual disability, visual tasks may be performed with the use of special aids and lighting.

The fact that a person has a visual impairment tells us nothing about what he or she is like, what he or she can do, or even how much he or she can see. Blindness is not debilitating. The general public's attitudes and prejudices are more likely to be a handicap than the visual disability itself.

Terms Related to Vision

Some potentially confusing terms are used to specify various levels of visual impairment. A brief overview of these terms follows:

Amblyopia (lazy eye). This common condition involves poor vision in an eye that did not develop normal sight during early childhood. Amblyopia affects 2 or 3 out of every 100 people (American Academy of Opthalmology, 1995).

Blindness. A person who is totally without the sense of vision or has only light perception is considered to be blind. Such a person must learn primarily through touch and hearing.

Farsightedness. This refers to a condition in which one can see objects at a distance more clearly than those near at hand.

Legal blindness. Visual acuity of 20/200 or worse in the best eye with correction is considered to be legal blindness. This means that an individual can read at 20 feet what a person with normal vision can read at 200 feet.

Low vision. People with low vision have limitations in distance vision but are able to see objects and materials within a few inches or feet.

Nearsightedness. This refers to a condition in which one can see distinctly at a short distance only.

Partially sighted. People who are partially sighted have a visual acuity greater than 20/200 but less than 20/70 in the best eye after correction. People who are partially sighted are able to use their vision as a primary source of learning.

Residual vision. Any usable remaining vision is called "residual vision." For example, if an individual can detect only light, he or she can use that ability to an advantage.

Tunnel vision. This describes the field of vision limited at its widest angle to 20 degrees or less. A normal field of vision is generally measured on a horizontal arc of 160 to 180 degrees. A person with severe tunnel vision is considered to be blind.

Issues that Persons with Visual Impairments May Face

- Accessibility
- Adaptable equipment
- Finding support
- Finding proper optical care
- Self-esteem
- Relationship and marriage issues
- Transportation and mobility
- Labeling
- Finding role models and mentors
- Dealing with other people's discomfort
- Restrictions on choice of job due to perceived limitations
- Feelings regarding the need for assistance
- Independent living

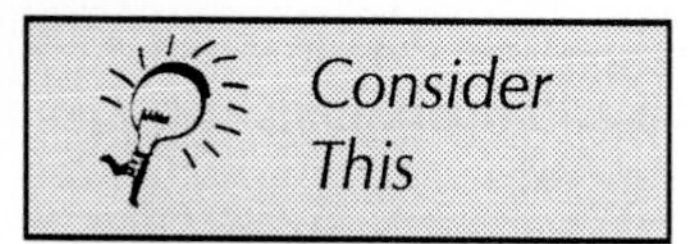

Always indicate the end of a conversation. This will avoid the embarrassment of leaving a person speaking when no one is actually there (American Foundation for the Blind, 1995).

Support Services for Vision Loss

People with visual disabilities use various types of adapted equipment to assist them at work, school, and throughout daily life. Some are quite ordinary, such as black felt tip markers that produce darker print, adjustable lamps to increase the amount of light and adjust its direction, large-type books, bifocals, contact lenses, magnifiers, tape recorders, friends, and guide dogs. More sophisticated technological aids include computers with speech output and enhanced screen images, Braille printers, and optical scanners and readers.

The local Division of Blind Services offers a wide range of information, from how to receive mobility training , how to obtain a guide dog, and where to find Braille newspapers or reader services. Public Radio, in many communities, also provides reader services by reading new books and the daily newspaper each day.

Interactions

Speak first and identify yourself. Let the person know who you are before he or she has to ask.

Use the person's first name. When in a group, using the person's first name, each time you speak to him or her, will make it clear who is being addressed.

Use a normal tone and volume of voice. People with visual impairments do not necessarily have trouble hearing, so speak naturally.

Be precise when describing something. Provide a reference point, such as "next to the door you came in" or "to your immediate left, about shoulder level."

Give tactile clues. Be aware that the person may rely on other senses to provide information. You can help to provide tactile clues: "Joan, let me take your hand and show you the skirt I just made."

Don't assume that help is needed. Ask if you can help, and then follow the lead of the person.

Guide persons who request assistance by allowing them to take your bent arm just above the elbow. Walk slightly ahead of the person you are guiding. Never grab people by the arm and push them forward (American Foundation for the Blind, 1995).

Feel free to use words that refer to vision during the course of conversations. Vision-oriented words such as *look*, *see*, and *beautiful* are perfectly acceptable. Making reference to colors, patterns, designs, and shapes is also perfectly acceptable (American Foundation for the Blind, 1995).

Jot down from memory three possible causes of visual impairments.

Conclusion

It is impossible to make generalizations about people with hearing or vision disabilities. Few people have perfect hearing or perfect vision, and we all rely upon various technological devices to extend the range of our senses. Individuals with sensory impairments experience the same feelings and emotions as everyone else. They also have the same potential for communicating their ideas, fulfilling their dreams, and living dynamic, interesting lives.

References

Ackerman, D. (1990). *A natural history of the senses.* New York: Random House.

American Academy of Opthalmology. (1995). What is amblyopia? In *Amblyopia FAQS.* [On-line.] Available: http://www.eyenet.org/public/faqs/amblyopia_faq.html

American Foundation for the Blind. (1995). Sensitivity to blindness or visual impairments? In *Information center.* [On-line.] Available: gopher://gopher.igc.apc.org.5005/00/info/general/sensitiv

Chopra, D. (1991). *Unconditional life.* New York: Bantam Books.

Idaho State University. (1996). Facts about children at educational risk with hearing problems. In *Department of Speech Pathology and Audiology Home Page.* [On-line.] Available: http://www.isu.edu/departments/spchpath/trai/facts.htm

LaPlante, M. P. (1991). The demographics of disability. In J. West (Ed.), *The Americans with Disabilities Act: From policy to practice* (pp. 55-80). New York: Milbank Memorial Fund.

Reynolds, M. C. & Birch, J. W. (1982). *Teaching exceptional children in all America's schools.* Reston, VA: Council for Exceptional Children.

Toufexis, A. (1991, August 5). Now hear this—If you can. *Time,* pp. 50-51.

Suggested Readings

Baldwin, S.C. (1993). *Pictures in the air: The story of the National Theatre of the Deaf.* Washington DC: Galludet University Press.

This book offers a window into the fascinating history of the theatre for the deaf and the individuals who helped to shape it.

Bienvenu, M. J. & Colonomos, B. (1993). *An introduction to deaf culture: Rules of social interaction.* Burtonsville, MD: Sign Media.

This is a fascinating exploration of the complex world of deaf culture.

Brady, F. B. (1994). *A singular view: The art of seeing with one eye.* Annapolis, MD: Frank B. Brady.

This book is a first-person account of living with monocular vision and finding ways to improve daily life.

Cohen, L. H. (1994). *Train go sorry: Inside a deaf world.* Boston: Houghton Mifflin.

This book is an intriguing look at the students and social condition of the Lexington School for the Deaf.

Keller, H. (1956). *The story of my life.* Boston: Houghton Mifflin.

In this fascinating autobiography, Keller describes the joy of learning to speak her first word.

Koestler, F. A. (1976). *The unseen minority: A social history of blindness in America.* New York: McKay.

This book traces the little-known history of individuals with visual disabilities in America.

Krebs, B. M. (1987). *Braille in Brief.* Louisville: American Printing House for the Blind.

This is an elementary guide to learning Braille. Includes charts of Braille characters and contractions.

Preston, P. M. (1994). *Mother father deaf: Living between sound and silence.* Cambridge: Harvard University Press.

This book is an account of growing up hearing but having two parents who cannot hear. A compelling examination of how deafness affects family relations.

Notes

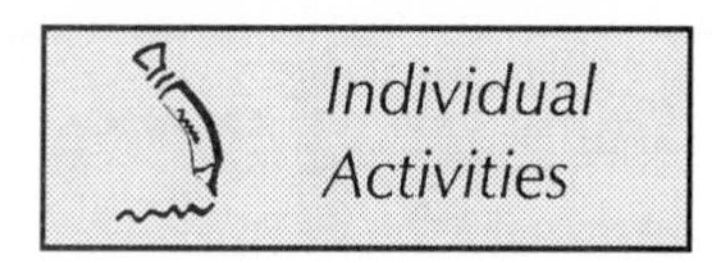

1. Watch an entire prime-time television program, either comedy or drama, with the sound off. Were you able to follow the show? What things proved helpful to your understanding? What things contributed to your confusion? Write your reactions below:

2. Blindfold yourself upon getting out of bed in the morning, then do everything you normally do to get ready for the day, including taking a shower and getting dressed. (Be careful!)

List three things that you found difficult, frustrating, or even frightening.

List three things you were able to do which surprised you.

1. Using only gestures and movements, help the members of your group guess the movie, book, or famous person you are describing.

Reactions:

2. Have your group divide into threes to participate in simulated community interactions with vision impairments. Assign roles:

Person #1: individual with vision impairment
Person #2: friend of person #1
Person #3: community worker

Scene One: fast food restaurant. Individual with vision impairment orders food and may have questions about selection, size of portions, etc.
Scene Two: asking for directions. Individual with vision impairment asks someone at the front desk for directions to a room on the third floor.

Discuss the communication between community worker and friend, bypassing the individual with vision impairment. Discuss the precision of the language that the worker uses. Examine the role of the friend as an intermediary.

Reactions:

Reflection Paper 14.1

Reflection Paper 14.1

"Sign language fluency should be required for high school graduation." Support or refute this policy.

Reflection Paper 14.2

Suggest proper actions and behaviors when interacting with a person with a visual disability.

Section III:

Perspectives on Diversity

Chapter 15 Family Perspectives

Chapter Sections

- Introduction
- The Family Learning Environment
- Types of Families
- Child Abuse
- Embracing Differences
- Terminology
- Support Services
- Interactions
- Conclusion
- Individual and Group Activities

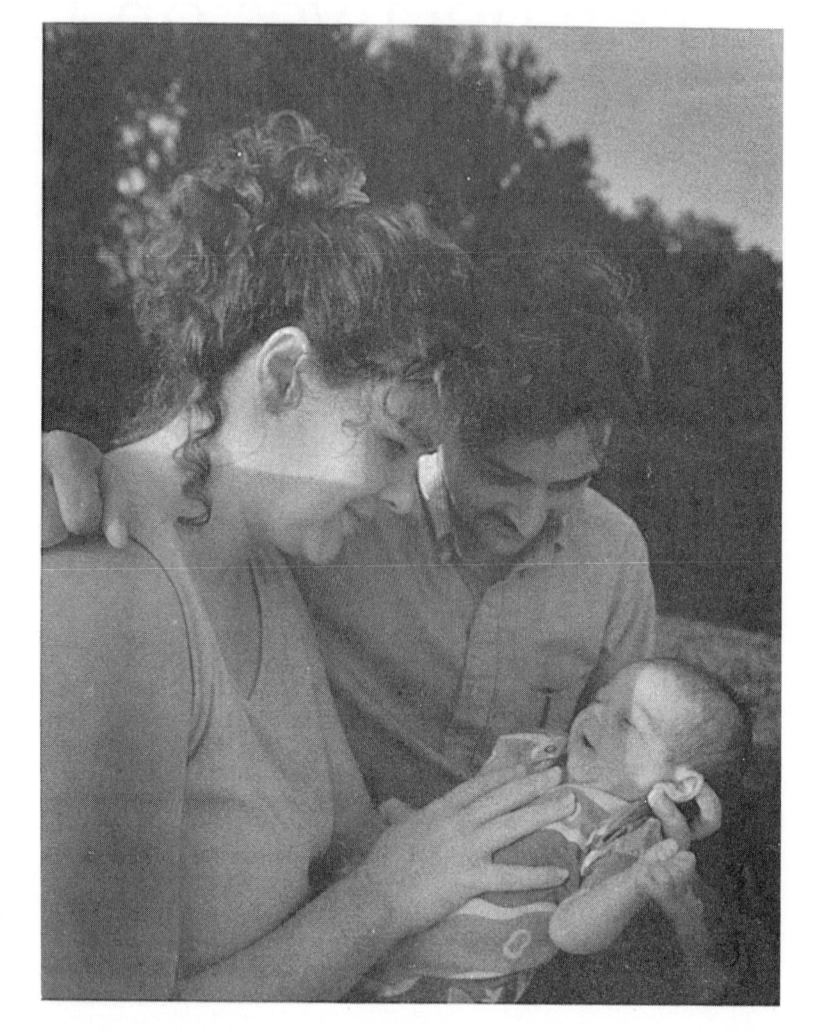

"No 'formula' for examining familes exists."
—Belinda D. Karge, educator

Objectives

- Identify three ways a family prepares its children for social interaction.
- Explain how child abuse affects everyone.
- Identify the most current definition of *family*.
- Explain the process family members may go through when accepting a member who stands out as being "different."

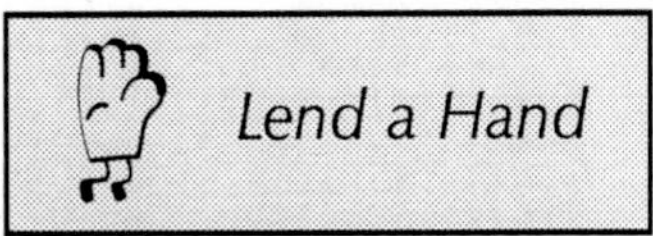

Demonstrate your respect and admiration for parents you come across in your daily life. Help them to survive and thrive by offering your assistance.

Introduction

The vital functions that families provide for their members are complex. They aren't things that one can summarize into a tidy list and post on the refrigerator door. Intricate and subtle connections, interactions, involvements, and interrelations take place which help all the members to develop and thrive (see Figure 15.1). Just as it is important for parents to support their children, it is equally important for parents to receive assistance, encouragement, admiration, respect, and affection from nearby relatives, friends, neighbors, fellow employees, and teachers. The quality of an individual's life is equally tied to family dynamics and to social interactions beyond the family.

Nancy Seufert-Barr, writing for the *United Nations Chronicle,* noted that "the family is a universal phenomenon—throughout the centuries and throughout the world" (Seufert-Barr, 1994, p. 1). Families are manifested in a wide variety of forms, however, and the idea of the family's role varies greatly among cultures.

Figure 15.1
Every member of a family develops and grows by participating in various interrelations.

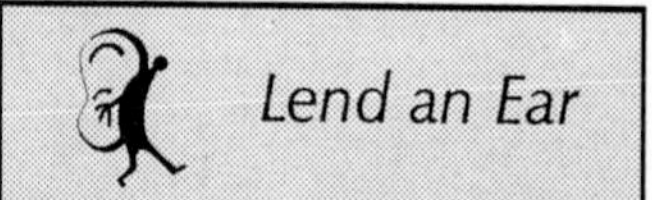

"The little world of childhood with its familiar surroundings is a model of the greater world. The more intensively the family has stamped its character upon the child, the more it will tend to feel and see its earlier miniature world again in the bigger world of adult life."
—Carl Gustav Jung, psychologist

There is no such thing as a "typical" American family. Families fit no standard size, socioeconomic level, country of origin, age breakdown, religion, race, or makeup by gender or sexual orientation. However it is defined, the family unit is responsible for providing basic necessities of life, including shelter, comfort, food, security, friendship, love, attention, nurturing, trust, and a sense of worth and dignity.

The Family Learning Environment

The most important learning environment for any child is the family. It is within the family that children learn what it is to be a person. It is there that they learn the emotional, behavioral, and value patterns which form their identities and set up their social interaction with the rest of the world (Dwinell & Baetz, 1993). What children learn is dependent upon factors such as what type of family they belong to, their racial heritage, ethnicity, socioeconomic status, language, and level of education, and the family's ability to prepare them to be a part of the larger culture. All these factors have a significant impact upon a child's success in life.

Government advisor on family policy Richard Weissbourd has pointed out that in many cases economic stresses in the family play a significant role in a child's academic life (1996). Weissbourd says that children from poor families may need to work to earn money for the family, care for adult relatives, or provide child care for a younger sibling. He cited a survey of high school students which showed that about 20 percent of students reported having missed days of school so that they could care for a family member or close friend. In 1992, according to Weissbourd, "12 percent of high school dropouts nationwide reported that they had to care for a family member and 11 percent said that they had to help financially support their families (these groups overlap, since respondents could answer yes to more than one question)" (Weissbourd, 1996, p. 27).

The educational level of the parent may have an even more significant effect upon the child than the family's socioeconomic status. Studies have shown that "children growing up with uneducated parents who are not poor are more likely to live in damaging conditions than are children who grow up with educated, poor parents. Children who grow up with an uneducated parent also have more school troubles than do children who grow up with poor parents. They are more likely to be in the bottom half of the class and to be retained a grade than are children who are poor only" (Weissbourd, 1996, p. 25).

The family unit is responsible for providing basic necessities. What basic necessities do your family or circle of friends provide for you? How is your family or circle of friends a source of attachment, identity, and identification for you?

Types of Families

Family structure in the past was largely based on one of two types, nuclear and extended (Seufert-Barr, 1994). Nuclear families usually are made up of just two generations, parents and their children. These include single parent and adoptive families. Extended families include several generations and may be built upon a social rather than a biological basis. Nontraditional families are becoming more and more common. By nontraditional we mean families based on cohabitation, same-gender, single parent, and reorganized, or step families (Seufert-Barr, 1994).

Historically, most families have been patriarchal, but with the industrial revolution came profound changes in the structure of the family. Urbanization caused different work and life styles which contributed to the dissolution of extended families, as well (Seufert-Barr, 1994). Until fairly recently, the traditional nuclear family was made up of a father, a mother, and children living together in the same home. The father was the primary wage-earner, and the mother stayed at home to

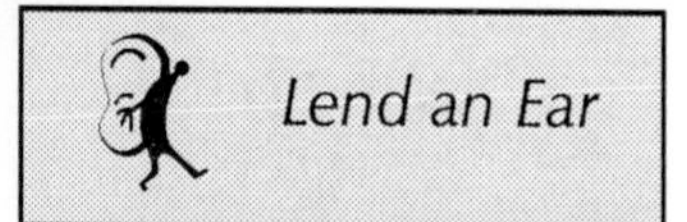

"Happy families are all alike; every unhappy family is unhappy in its own way." — Leo Nikolaevich Tolstoi, author

care for the children. Nuclear families make up less than 10 percent of American families today (Cushner, McClelland, & Safford, 1996).

Our definition of family continues to change as society changes. New definitions include adoptive, foster, and migrant families as well. Families may be made up of stepparents and siblings, relatives, single parents, or a parent and a partner. Children may be raised by grandparents, aunts, uncles, siblings, neighbors, or friends. We define families as the primary source of the child's care, support, and sense of belonging, whatever the demographic makeup of the household.

According to Weissbourd (1996), as many as half the children in the United States born in the first part of the 1990s will be children of divorce. Children may act out, test adults to see whether they will be abandoned, or deliberately cause trouble in the hope that it will draw their parents back together. They may be withdrawn or depressed, or too absorbed with personal matters to concentrate on anything else (Weissbourd, 1996).

Migrant families may experience particular problems of their own, often due to adjusting to a new culture and language or dialect as well as to frequent moves. They may also be living at a low socioeconomic level, and could be without the health or social welfare resources others might have.

Child Abuse

Circumstances Which Could Lead to Child Abuse:

- Marital discord
- Alcoholism
- Drug addiction
- Immature parents
- Criminal environment
- Illiteracy
- Family values in conflict with societal values

While ideally the family is a refuge and source of strength, comfort, and protection for its members, far too often children suffer from abuse, violence, and neglect within the home. Child abuse can be defined as behavior which negatively affects a child's physical or emotional health. Sexual abuse ranges from nonphysical to violent. Abuse takes place in families of every category of wealth, education, race, geographic location, and ethnicity. No one is necessarily immune.

The problems of abuse affect every person in the United States. Even if no child you know personally is a victim of abuse, the climate is pervasive, and it colors our worldview. As a society, we cannot ignore the reality of the family crises so many our children experience, because these children bring the effects of abuse everywhere they go and we may be called upon to respond in a variety of ways.

Aside from the physical consequences of abuse, other effects might also show themselves. A child who has been abused may suffer from depression. He or she may be unable to focus, and may be too preoccupied by problems at home to focus on academic or other matters (Greenberg, 1994).

Social researchers now believe that one in three or four girls will be sexually abused by age eighteen. One in eight to ten boys will be victims of abuse, with the mean age between six and nine (Dumas, 1992). According to the National Center on Child Abuse Prevention, reported cases of child abuse rose 147 percent between 1979 and 1989, and cases continue to rise. Sexual abuse accounted for 15 to 16 percent of those reports (Dumas, 1992).

In addition to direct correlations to lack of success in school, children who have suffered abuse often lose their sense of self-esteem. Many children who become runaways, teenage prostitutes, prison inmates, and suicides have been victims of child abuse. Though some children are able to recover from their experiences, others are affected for the rest of their lives (Greenberg, 1994).

If you have reason to believe that a child you know is now being or has in the past been abused, it is important for you to take appropriate action. Be sure to follow the laws of your state regarding child abuse. In some states you are required to report suspected child abuse to the Department of Public Welfare-Child Protective Services Division. Consult the Yellow Pages or a local law enforcement agency for the number. You can

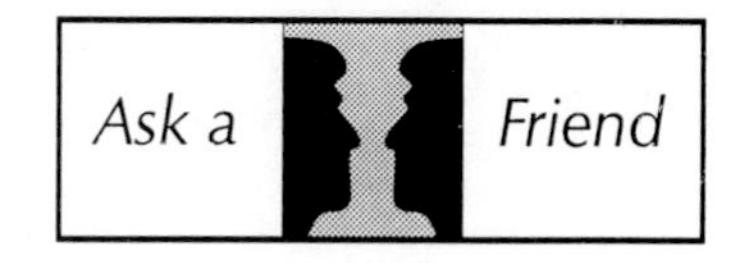

How widespread is the problem of abuse? What percentage of children would you assume to have had some experience of sexual or physical abuse?

Possible Signs of Child Abuse or Neglect:

- Child is overly compliant
- Child is withdrawn and passive
- Child is uncommunicative
- Child has an unexplained injury
- Child is dirty, smells, has bad teeth, missing hair or lice
- Child is unusually fearful
- Child is thin, emaciated, and constantly tired, showing possible evidence of malnutrition and dehydration

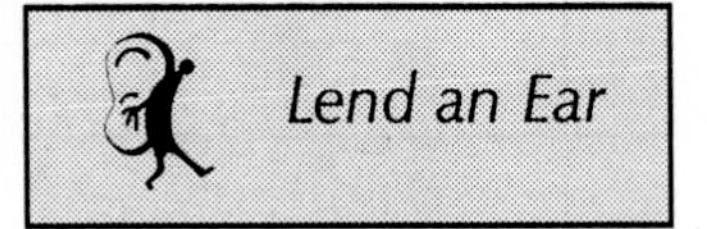

"The family only represents one aspect, however important an aspect, of a human being's functions and activities.... A life is beautiful and ideal, or the reverse, only when we have taken into our consideration the social as well as the family relationship."
—Havelock Ellis, psychologist and writer

file a confidential report, and in most states an investigation is required to be initiated within 24 hours of the report. To take no action may be inviting disaster for the child and may, depending upon the laws in your state, be illegal.

Embracing Differences

A family may face additional challenges or frustrations when one of its members stands out as being "different" in some way. One member may join a different political party, may exhibit mental illness, may choose a different religion, may behave differently, may have a different sexual orientation, or may be born with or acquire a disability. Still, there are many common ties that can unite family members as they learn to embrace diversity and eventually benefit from it.

Turnbull & Turnbull (1990) explained that the process of accepting an exceptional family member may be similar to that of accepting the death of a loved one. However, if long-term planning, empowerment and advocacy, community integration, and financial planning are taken into consideration, the grief can be turned into something positive. Feelings of grief may arise in times of stress. Families may grieve for the child they have and for the child that might have been. They may grieve for relationships that are permanently changed. Or they may grieve for themselves and for some imagined ideal life. The process may include any of the following emotions: shock, depression, isolation, denial, guilt, shame, anger, fear, uncertainty, and acceptance.

No family can be without diversity because every individual is unique. When the members of the family realize this fact, they tend to embrace differences and to view them as enriching opportunities for growth rather than as challenges or problems. However, it is often the case that some members of a family feel strongly about their traditional religion, ethnicity, politics, or lifestyle. These individuals may go through a process of adjustment and acceptance which can best be aided by education.

The reason that diversity is a positive aspect of society is that fear of the unfamiliar inhibits the exchange of ideas and the cooperation which promotes enlightenment and the advance of humankind. The experience of diversity among those whom we love best motivates us to give up our fear and to move forward. Therefore, all of society benefits from our familial feeling.

Make a list of all the diverse qualities of the people in your family or circle of friends.

Terminology

The following terms and concepts need to be considered when thinking about the family's role in society:

Child abuse. This is any behavior that hurts a child physically or emotionally.

Circle of friends. This refers to the many people whose lives overlap ours. We have an inner circle of relatives, bonded by family ties. A second circle includes close friends. Third, we have casual acquaintances. Finally, we have those who are paid to be in our lives, such as our teachers, doctors, and employers.

Disabled abuse. This refers to abuse of individuals with disabilities, including neglect.

Domestic abuse. This refers to the abuse of one family member by another.

Emotional abuse. This includes using words or withholding affection to hurt someone. Emotional abuse often involves making someone feel unloved, unwanted, insecure, and unworthy.

Do you consider the "latchkey" phenomenon to be a form of child abuse?

Extended families. This is a family of several generations living together in the home, possibly including aunts and uncles.

Latchkey kids. This refers to children who spend longer than three hours at home every day without adult supervision.

Migrant family. This is a family which has relocated to a new country and faces the burden of cultural change along with other adjustments. It may also be a family which faces many changes resulting from constant movement due to employment opportunities.

Neglect. This involves failing to give a child, person with a disability, or elderly adult proper shelter, food, clothing, or health care (Greenberg, 1994).

Nontraditional family. Families made up of adults who are cohabiting, same-gender partners, single parents, and their children are called nontraditional, though they are rapidly growing in number.

Nuclear family. This is a family consisting of two generations in the home, parents and their children. Nuclear families include one-parent and adoptive families.

Reorganized family. This is a family which has been created through marriage, remarriage, or cohabitation of people who had children by former partners.

Sexual abuse. This type of behavior ranges from obscene phone calls and indecent exposure to rape (Greenberg, 1994).

Throwaways. These are children who are either forced out of their homes or abandoned by their parents (Greenberg, 1994).

Verbal abuse. This involves using cruel language and insults to hurt someone (Greenberg, 1994).

Lend an Ear

"Americans founded the Society for the Prevention of Cruelty to Animals before we extended the same protections to our children."
—Belinda D. Karge, educator

Support Services

Support groups consisting of parents and family members with common experiences and situations are usually the most successful at providing assistance and sharing information. Parents and family members learn that they are not alone, receive useful information, share concerns and fears, and give each other the benefit of their insights (Lessen, 1991).

Interactions

Remember that your knowledge of other people's families is limited. You probably know very little of their private situations, just as they probably know very little of your own. While your assessments must be based on your own observations of behavior, keep in mind that family dynamics are usually highly complex.

Be aware of signs of abuse. These may include: spontaneous statements made by the child; precocious sex play or talk, infantile or aggressive behavior, uncharacteristic cheerfulness, secretiveness or vigilance, and sudden changes in mood or action. Be aware that any of these signs may also occur where there is no abuse, so be alert but don't to jump to conclusions (Dumas, 1992).

When speaking with parents, always have something positive to say. Raising a child is a difficult task, and parents need support and encouragement.

What are some possible reasons why reports of abuse and neglect have tripled in the last decade?

Conclusion

We must stay aware that when we feel powerless to fix problems, sometimes we stop seeing them altogether. Left unattended, however, these problems grow into devastating social failure, which leads to such things as alienation, dropping out of school, drug abuse, and long-term welfare dependency. While we may not be able to make dramatic changes, we can fix upon small deeds which make a significant difference in a person's life and prospects for a positive future. We can help to secure glasses for a child with vision impairment, find school supplies for a child who is in need, and stay aware of the pervasive threats to children's learning (Weissbourd, 1996).

Though families are the source and the center of children's lives, we must also remember that children do not exist in a vacuum. As Weissbourd (1996) pointed out, children "are not merely products of advantaged or disadvantaged home environments—they are not putty in their parents' hands—they are powerfully shaped by a vast array of circumstances in the larger world" (p. 27). As a society, we must go beyond the stereotypes of race, class, and family income if we are ever to effectively address the problems which so profoundly afflict our children.

References

Cushner, K., McClelland, A., & Safford, P. (1996). *Human diversity in education: An integrative approach.* New York: McGraw-Hill.

Dumas, L. S. (1992). *Talking with your child about a troubled world.* New York: Fawcett Columbine.

Dwinell, L. & Baetz, R. (1993). *We did the best we could: How to create healing between the generations.* Deerfield Beach, FL: Health Communications, Inc.

Greenberg, K. (1994). *Family abuse: Why do people hurt each other?* New York: Twenty-first Century Books.

Lessen, E. (1991). *Exceptional persons in society.* Needham: MA: Simon and Schuster.

Seufert-Barr, N. (1994). *Families around the world: Universal in their diversity.* UN Chronicle, 31, 46.

Turnbull, A. P. & Turnbull, H. R. (1990). *Families, professionals, and exceptionality: A special partnership.* Columbus, OH: Merrill.

Weissbourd, R. (1996). *The vulnerable child: What really hurts America's children and what we can do about it.* Reading, MA: Addison-Wesley Publishing Company.

Suggested Readings

Hite, S. (1994). *The Hite report on the family: Growing up under patriarchy.* New York: Grove Press.

This book analyzes the changing shape of family life and seeks to legitimize the infinite ways that people live as "families," whether as single parents, same-sex partners, in traditional family groups, or alone.

Louv, R. (1990). *Childhood's future.* New York: Anchor Books.

This is a penetrating look at American children and their families.

Powell, T.H. & Ogle, P.A. (1985). *Brothers and sisters: A special part of exceptional families*. Baltimore: Paul H. Brooks.

This booklet suggests techniques and services that will help siblings of children with disabilities better understand the feelings and circumstances which surround the experience of growing up with a brother or sister with an exceptionality.

Rose, H. W. (1987). *Something's wrong with my child.* Springfield, IL: Charles C. Thomas.

This is a personal account of a mother's experience living with a child with a disability.

Weston, K. (1991). *Families we choose.* NY: Columbia University Press.

This book examines the subject of lesbian and gay kinships.

Individual Activities

1. In the space below, draw your own diagram of your circle of friends and family. Who are the most important people around you and how do they function in your life?

2. Each person in a family plays more than one role and participates in more than one relationship. For instance, you may be a mother, a sister, a sister-in-law, a daughter, a wife, a cousin, a peace-maker, and a friend all at the same time. How many roles do you play in your family, and what is expected of you in each role?

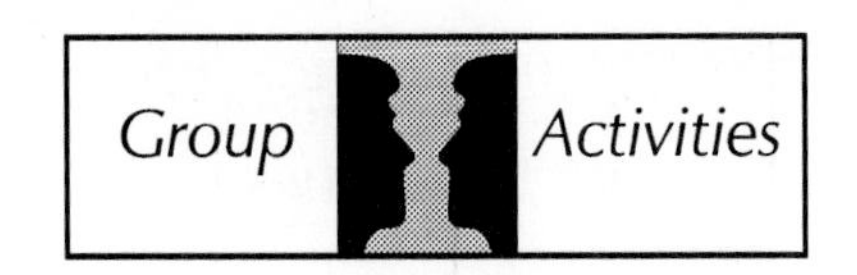

Divide yourselves into groups of 4 or 5 students. This is your new family! Choose a mom, dad, brothers, and sisters. Now choose one of the scenarios below.

Scenario A: Your sister Julie is not with you at this time. She has run away from home because she felt different from the rest of the family and didn't know how to explain her feelings. Discuss your feelings about Julie's situation. What will you say to her if/when you see her again?

Reactions:

Scenario B: The mom and dad have different political viewpoints and are 20 years apart in age. One of the children is married to someone of a different race. Another of the children has chosen a religion different from the parents'. One grandparent has either a disability, a different sexual orientation, or speaks a different language. The family is spending time together over a holiday. Imagine that you are all at the dinner table and one of you starts talking about the ties that bind the family together. What do you all have in common that keeps you so close?

Reactions:

Reflection Paper 15.1

"The experience of diversity among those whom we love best motivates us to give up our fear and to move forward." Do you agree with this statement? Why or why not?

Reflection Paper 15.2

The number of single-parent families has soared in recent years. Since this is a trend that isn't likely to go away overnight, what are some specific things society can do to help ease the difficulties of single-parent family life?

Notes

Chapter 16 Educational Perspectives

Chapter Sections

- Introduction
- Quality Education for All
- Modern Risk Factors in Students' Lives
- Current Controversial Issues in Education
- Education Models
- Racial Considerations
- Gender Considerations
- Disability Considerations
- Language Considerations
- Sexual Orientation Considerations
- Other Considerations
- Conclusion
- Individual and Group Activities

"Education is not just another consumer item. It is the bedrock of our democracy"
—Mary Hatwood Futrell, educator

Objectives

- Identify modern risk factors affecting students' academic success.
- Distinguish three general systems of schooling.
- Discuss current controversial issues in education.
- Explain the six fundamental policies in PL 101-476.
- Define the requirements for fair, appropriate education for all students.

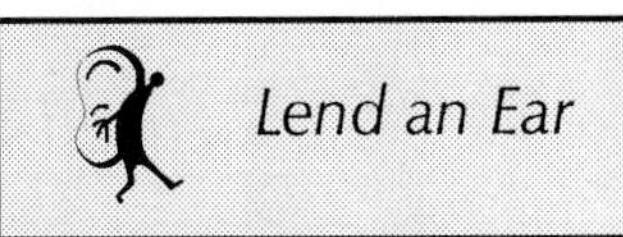

"Schools are complex places. The more closely they are examined, the more complex they become."
—Kathryn Whitmore, professor

Introduction

Schools and classrooms do not stand apart from the larger society, but are, in fact, microcosms of the society. When we as children, adolescents, or adults enter a classroom, we come as we are. Our race, age, religion, culture, sexual orientation, physical health, language, and ability to learn in a prescribed way do not alter from what they are in our lives away from school. If the reality of who we are is ignored or denied, then it will be difficult for us to learn much beyond fear and cynicism (see Figure 16.1).

While most people would agree that education helps to develop greater independence and more productive ways of living, they might well disagree as to who should be educated, how children, adolescents, and adults can most effectively and efficiently be educated, and even as to exactly what we mean by "education." Many non-education majors will read and use this book. Therefore, it is not our purpose to suggest educational interventions or teaching strategies. However, everyone should have a good understanding of our school system and how it seeks to

Figure 16.1
Every child is unique, and that uniqueness must be acknowledged for learning to occur.

What are some factors which may influence the quality of educational programs?

serve our diverse society. This chapter should provide an overview of school policies relating to diversity, specific services provided by public and private schools due to federal laws and policies, and considerations which must be made to ensure that all children receive a free and *appropriate* education.

Quality Education For All

The education offered in America's schools has been designed to provide everyone with equal opportunities to learn academic subjects as well as the social and basic vocational skills needed for success after high school. Throughout history, however, certain children who are not citizens, who are members of a minority race, gender, or sexual orientation, or who are from an ethnic or religious minority have often not had equal educational opportunities. In addition, the quality of educational programs may be influenced by geography, the economy of the specific community, the philosophy of the state or local government and school board which administers the program, and by the education and background of the teachers and administrators in the school system.

There are other influences as well, such as whether the program is provided by the public school system or by a private entity such as a church or private charter school, or whether it is administered by parents or others interested in a specific philosophy. In an effort to make individual schools more responsive to the needs of their community, many public school districts have decentralized and have created school advisory councils, made up of teachers, parents, school administrators, and members of the community. While this appears to be generally successful, it also provides an opportunity for a few individuals, who may not have a full understanding of the needs of all individuals, to provide direction for an entire school. If a particular student or group of students is not represented or considered by the school advisory council, their needs may not be met.

It is not only important that you have a better understanding of special considerations in schools for children who are diverse, but it is also important to recognize that you may someday be able to influence the quality of school programming. You may one day be a parent or relative of a child who requires special school programming. This chapter should aid you in interacting with educational professionals and in requesting specific services which may be beneficial to someone you know. You should also be better able to positively influence the educational philosophy and programming in your community. No doubt many readers of this textbook will become community leaders. You may serve on your local school board, be on a charter school board, be a member of a school advisory council, or be a board member in a church or synagogue where a private school is managed. Your study of diverse populations and your attention to diverse learners may offer you insights and direction in providing leadership in education.

Jot down twelve modern risk factors affecting students' academic success. Then check your answers on the next page.

Modern Risk Factors in Students' Lives

Most educational experts tend to agree that home life and parental involvement are crucial to a child's achievement in school. However, teachers are finding that an increasing number of their pupils do not have a supportive home life, either living in broken homes or in homes in which both parents work and have little time for the needs of their children. As Landes (1994) observed, some teachers are over-whelmed with obligations to provide emotional support and values to children to who do not receive them from their parents. "In some classrooms, the need to teach children the basic skills of living (brushing one's teeth, for example) has supplanted academic instruction" (p.102).

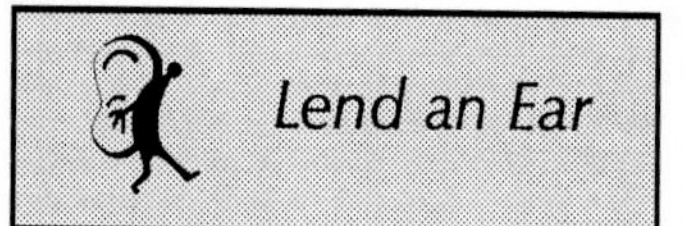

"The school must mirror a society that is ready to embrace all children, so they can be educated together and learn to value one another as unique individuals."
—Belinda Karge, professor

In addition to lack of parental involvement, other modern risk factors affecting students' academic success (adapted from Landes, 1994) include:

- Home alone more than three hours on weekdays
- Involvement in pregnancy
- Use of alcohol or drugs by student and/or family
- Physical or sexual abuse
- Excessive absences
- Failure of courses
- Low self-esteem
- Sickness or death of parent
- Parent's loss of job
- Death of friend or sibling
- Illness
- Parent's negative attitude
- Limited English proficiency
- Broken home
- Move frequently
- Parents divorced recently

Students with two or more risk factors are twice as likely to be in the lowest academic achievement quartile and are over six times as likely to think that they will not finish high school (Landes, 1994). Clearly, the daily problems that children are confronted with outside of school directly affect their academic performance. Interventions must begin in the home—one family at a time—if we ever hope to see a significant reduction of such problems in the classroom.

Current Controversial Issues in Education

As our society continues to embrace diversity, our schools often find themselves in delicate positions as they struggle to bring multicultural perspectives and experiences to the classroom. In the past, school curricula centered on the cultural aspects of Western civilization, to the exclusion of other cultures and languages. Histories are now being rewritten from a global perspective to include ethnic groups that had previously been excluded. Curricula are constantly being revised to include the interests of more categories of diversity. Yet even as our schools attempt to stay abreast of current trends, they frequently come under fire by groups of parents who disagree about how to teach certain controversial subjects. These include:

What is the best way to solve the controversy surrounding sex education? School prayer? Ethnic history?

- Ethnic history and literature
- Sexual orientation
- Sex education
- AIDS education
- Values
- Basic skills of living
- Gender studies
- Minority studies
- Observance of religious holidays
- School prayer
- Learning styles of different ethnic groups
- Bias in standardized testing
- Bilingual education
- Inclusion

Though the issues themselves will change over time, our educational system will no doubt continue to find itself in the firing line as our society redefines its attitudes about cultural pluralism and our schools seek to reform their goals and objectives concerning human diversity.

Education Models

Because of our diverse population, throughout the history of the United States there has been a need to serve a variety of social and cultural groups. Historically, much of our country's education was directed by church groups. In colonial days education was often dominated by the need to teach children both academics and the morals which agreed with a particular

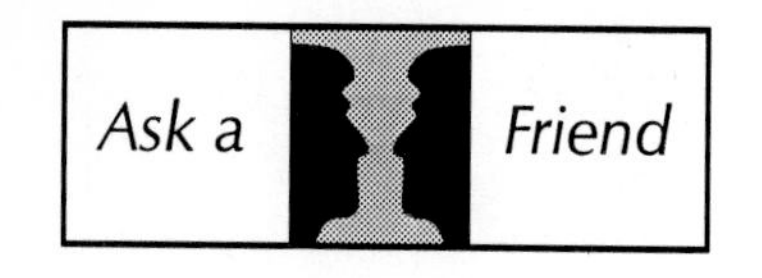

What are the three general systems of schooling in the United States.

church doctrine. However, in the 1830's American schooling began to be for and by the public to a greater extent, so that children of different religious denominations could be educated together. This common school movement was designed to provide an education which supported the democratic form of government (Cushner, 1996).

Today there are three general systems of schooling in the United States: public, private/parochial, and alternative. Each system offers its own unique benefits.

Public school. The vast majority of students in the United States attend public schools, which are regulated and funded by state, local, and federal governments. Curricula is designed and chosen which seeks to meet the needs of individual children within a structure which recognizes and respects diverse cultural, social, academic, physical, and language differences. Parents choose public schooling for socio-economic reasons—it is supported by tax dollars—and also for the excellent programs and facilities public schools frequently offer which may not be feasible for private schools with smaller student populations and more limited budgets. Another reason parents may prefer public schooling is that it offers social interaction with a broader range of students and allows for the experience of learning within an environment which is more likely to be composed of persons who are diverse.

Private/Parochial schools. Private and parochial schools may also receive some government funds, but they are primarily tuition-driven. They also have to meet state standards for curriculum, but their individual philosophies and goals are focused on a particular point of view. This perspective is religious in the case of parochial and church schools, and philosophical in the case of many other private schools. For example, for the past half-century throughout the United States and

Europe, Montessori schools have flourished. These are small private schools based on the educational philosophy of an Italian educator named Maria Montessori. She developed her teaching methods for children with disabilities, but as her books were widely read around the world, people began to apply her ideas in all classrooms. Other private schools stress more traditional teaching methods and use such models as the old Latin schools.

Though private and parochial schools are largely privately funded, many seek to make their student body more diverse by offering scholarships to disadvantaged students. Free, universal schooling is at the core of our requirement as a democratic society to have an educated citizenry. However, we are enriched by the diversity of avenues to education, and parochial and private schools serve the important function of preserving and fostering diverse and distinct cultural and religious points of view.

Lend an Ear

"Educational institutions, like any other segment of American institutional life, must *embrace diversity*. Why? Because it is the rational, logical, and intellectual thing to do given the changes in our society. To do otherwise is to continue to insulate ourselves from the very society we are charged with serving." — Maricopa Community College

Alternative schools. The reasons people have for choosing an alternative method of schooling for their children are as varied as the individuals involved. Thousands of parents in every state choose to teach their children at home. Some of them are interested in meeting the special physical, emotional, or academic needs their children may have. Many families find that the flexibility of home schooling allows for family travel, intensive study, and other educational experiences which traditional schooling inhibits. Home schoolers have many gifted students among their ranks, who may use their control over scheduling to devote time to music, art, or special academic interests. Many children with physical disabilities or health issues also profit from the individualized environment and one-to-one instruction which home schooling can provide.

Home schooling has a long tradition in America. All of the founding fathers of our country, including George Washington, John Adams, and Thomas Jefferson, were taught at home. In more recent times, Supreme Court Justice Sandra Day O'Connor was taught at home as well. Since home schooling has been legal in most states for over 20 years now, there are statistics showing that students typically achieve at about two grade levels above average, that they are generally socially active, and that a high percentage goes on to successfully complete college.

Racial Considerations

Political events have greatly influenced American views on diversity in schools. Prior to the Civil War, for instance, many American communities did not offer public education to African-American children. In fact, it was illegal to provide public supported education to children of slaves throughout the south (Cushner, 1996). Though the Civil War precipitated a long journey toward racial equality in schools, many educational and governmental leaders still question whether equal educational opportunities exist even now.

Segregated schools were the norm throughout much of the United States until the passage of the Fourteenth Amendment in 1868, which defined and guaranteed the rights of all citizens. When civil rights became a federal responsibility, school leaders began to recognize that federal funding for state schools would be lost if the rights of all children were not honored.

It was not until the Supreme Court ruled in 1954, in the now famous decision known as *Brown v. The Board of Education of Topeka,* that African-American and Caucasian children could not be served in separate schools, since separate implied that the schools were not equal. The courts of the United States have continued to deal with segregation issues in schools, but there are still areas across the country where equality of opportunity in education does not exist for all races.

Gender Considerations

It wasn't until the passage of Amendment XIX of the United States Constitution in 1920, that women were given the right to vote. Until that time, women did not enjoy equality with men in most segments of society, including schools. Although females attended public schools, our ancestors generally thought that the role of the woman was to raise a family and to be a homemaker. Therefore, many believed, it was not necessary to provide quality education for girls. Lower academic expectations of girls by teachers and parents, along with gender stereotyping in employment, negatively affected quality education in such areas as mathematics and science.

"Major shifts in values over the last three decades have broadened the educational prospects for girls in the United States" (Garcia, 1994, p. 48). However, it wasn't until 1963 that the U.S. Congress passed the Vocational Education Act which included provisions for equal educational opportunities for male and female students. In spite of these legal remedies, female students are still sometimes struggling for admission to certain private schools and even for equality is such school related activities as athletics.

Disability Considerations

Educational historians often begin discussions about schooling for children with disabilities with an overview of The Wild Boy of Aveyron. This story illustrates that a child, discovered in the woods of France in 1799 and raised by animals, was able to learn skills for daily living through the use of specially designed teaching techniques. Jean Itard, a French physician who taught the child and developed specialized and individualized educational plans, is considered to be the Father of Special Education (Smith & Luckasson, 1992).

What are the four main purposes of IDEA?

In early America, most children who had mental retardation or severe physical disabilities were routinely removed from the family and educated in large state-run institutions. Children who were blind and deaf also were offered specialized programs in institutional settings. At that time, institutionalization was thought to be best by many early leaders in special education; this of course is not the case today.

The first non institutional program in the United States for children who were deaf was founded in 1817 by Thomas Gallaudet; Gallaudet University, in Washington D.C., which specializes in college programming for persons who are deaf, is named after this early special education pioneer. Samuel Howe founded the first school for children with severe visual impairments, now called the Perkins Institute, in 1832 (Smith & Luckasson, 1992).

Public school education programs for children with disabilities began in 1878 when segregated special education classes were begun in Cleveland (Kanner, 1964). Separate classes and institutional programs, although they underwent many changes and improvements over the years, were the norm for children with disabilities until the 1970's. Many children with disabilities remained at home and did not receive formal or public supported education until passage of The Education for All Handicapped Children Act, Public Law 94-142, in 1975. Now called the Individuals with Disabilities Education Act (IDEA), this legislation has four major purposes, including:

- to guarantee the availability of special education programming to children and youth with disabilities who require it
- to assure fairness and appropriateness in decision making about providing special education to children with disabilities
- to establish clear management and auditing requirements and procedures regarding special

education at all levels of government
- to financially assist the efforts of state and local government through the use of federal funds

What does "mainstreaming" mean?

While most students with disabilities have been in separate special education classes in the past, placement of students with disabilities is now, by law, required to be in the least restrictive environment. That means that every child, regardless of the type or severity of his or her disability, should be in the most normal or regular school setting possible where he or she can be successful. This placement option has been called mainstreaming. The current preference among many educators is to place most children with disabilities in the regular education classroom and to provide whatever services are needed in that setting so that the children are successful in that environment. This model of serving children with disabilities in regular settings is called inclusion.

Students with disabilities are also required to have an individualized education plan (IEP). The IEP is a locally generated written document designed by a team of people interested in programming the child for success in school. Once the assessments have been completed, a team meeting is scheduled. The participants at the IEP meeting must minimally include:

- The school representative who will provide or supervise the provision of special education
- Child's teacher or teachers
- Parent(s) or guardian(s)
- The child (when appropriate)
- Other individuals at the discretion of parents/ committee

The team of experts contribute interdependently to each child's individualized program. Nationwide, IEP forms contain several standard elements:

- The child's present level of educational performance
- A statement of annual goals and short-term objectives

- Related services
- Percent of time in general education
- Beginning and ending dates for special education services
- An annual evaluation plan

Beginning no later than age 16, IEP's must include a statement of the "transition services" the student will need before leaving the school setting based on a post-school timetable. Transition services promote the student's movement from school to post-school activities (including post-secondary education, vocational training, integrated employment, independent living, or community participation) and must include instruction, community experiences, development of employment, and other post-school living objectives. In some cases, acquisition of daily living skills and functional vocational evaluation are requirements in the transition services.

Language Considerations

Throughout the history of the United States, those from other lands have immigrated in large numbers to America. Imagine the difficult adjustment which must be made by the child who arrives in a classroom that bears no resemblance to the educational program of home. These children are often fearful, anxious, disoriented, and may undergo culture shock (Grossman, 1995). Often teachers have no training in dealing with the needs of children who have arrived in their classrooms speaking no English and not understanding expectations and rules.

Bilingual programs were begun in the late 1960's to provide appropriate education for children whose first language was not English. Though their merits are widely debated, bilingual programs, which include the study of historical and ethnic aspects of other cultures, are prevalent throughout America's schools. Many years ago, government and school leaders believed in the 'Melting Pot' philosophy which stressed that all children would become acculturated and similar. Today the emphasis in education is for all children to learn basic skills while retaining their own cultural heritage.

Sexual Orientation Considerations

Students who are gay or lesbian are still an unprotected and neglected minority in today's schools. While laws have provided equal educational opportunities for all other groups within our society, those of a minority sexual orientation still have no legal protection. Children who are homosexual rarely find school reference materials on gay issues, hear about gay authors, historical figures, or sports heroes in their classes, or are offered curriculum which includes recognition that not all children and adults are straight (Schwartz, 1996).

While expressing the need for all government supported education to include facts about homosexuality in school curriculum, Andrew Sullivan, editor of the New Republic, has vividly pointed out the political difficulty in bringing this about (1995). Change seems to be underway with this last minority to be considered in education. In a landmark case, student Jamie Nabozny sued three school principals and the Ashland, Wisconsin School District for three years of ignoring his pleas for help from being harassed and physically abused by other students due to his being gay. On November 19, 1996, the court found the principals and school district to be responsible. The school district agreed to a $926,000 settlement (Price, 1996). This case sends a clear message to school officials: Treat all students with respect and protect all students from hate-filled behavior or face the possibility of severe legal consequences.

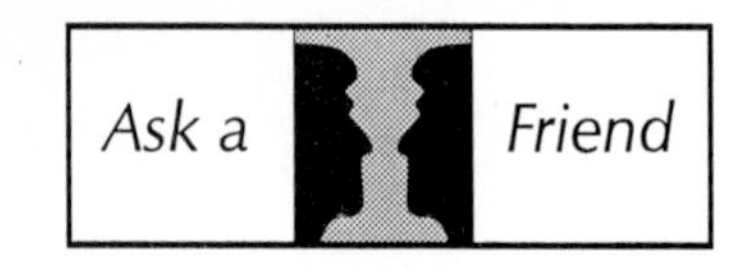

What are the six fundamental policies in IDEA?

Other Considerations

There are six fundamental policies in The Individuals with Disabilities Education Act (PL 101-476). As you read these policies, consider how applicable they are for all children in our schools.

Zero reject. Specially designed, free instruction will be available to all children with disabilities and no one (zero) can be rejected. "Free" means the education and related services will be provided at no cost to the person or to his or her parents or guardian. If a public education agency decides that a public or private residential program is appropriate, this type of placement (including nonmedical care and room and board) shall be provided at no cost.

Nondiscriminatory evaluation. Appropriately designed evaluation procedures must be followed. Each child must be tested in his or her native language (including such primary communication modes as Braille and sign language), and each must receive an individualized plan based on the results of the testing. The child cannot be placed in a program based upon a single assessment or the judgment of a single individual.

Due process. States must guarantee procedural safeguard mechanisms for children and their parents. For example, parents must have an opportunity to examine all relevant records. Prior written notice to parents and parental consent is required when any action respecting special education is proposed. Children and parents have the right to a formal, due process hearing if desired.

Parental participation. Parents must be involved in shared decision making and must provide consent to assessments and placements in special education services. Information about the child and family must remain confidential.

Least Restrictive Environment (LRE). Children and youth with disabilities must be educated with children who do not have disabilities to the "maximum extent appropriate." The LRE principle recognizes that a continuum of services and placements must be available to meet the child's needs.

Individualized Education Plan (IEP). The IEP is the management tool used to monitor the maximum LRE and therefore shall be applied within the framework of meeting the "unique needs" of each child.

Wouldn't these policies be appropriate for children with different languages? For those who are gay or lesbian? For children with different racial, cultural, or religious backgrounds? While special education teachers are specifically trained to respond to the individual learning styles of children with disabilities, perhaps the notion of a free and appropriate education for all children should truly include all children.

Following are some other issues to consider regarding the principle of a solid, fair, and appropriate education for all children.

Curricula. Colleges of education, where most teachers are taught, should provide a curriculum which is inclusive of all children. Future teachers need to know every child is "diverse" in one way or another, and that every child deserves to be treated fairly. Expecting a male child to be better in math is unfair to both males and females. Ignoring cultural feelings and customs is unfair to the culturally different child and impoverishes everyone. Excluding the child with a disability because the teacher 'thinks' the child can't succeed denies to the child with a disability the opportunity to succeed.

Learning styles. Children don't all learn the same way. If you were about to learn a complex task, such as how to assemble a bicycle, and you were given the opportunity to select from the following learning options, which one would you choose?

- read a book on the topic
- watch a film which shows the assembly
- read the directions and try
- watch someone else do the assembly
- participate in the assembly with a coach
- surf the web to find hints on the assembly
- hear a lecture on how to build the bicycle
- participate in a group which works together

Different people would choose different learning options depending upon their preferred learning style. Yet much of education, especially at the high school and college levels, is provided via the lecture method. Most probably that was not your preferred selection for the bicycle assembly operation, yet you are probably often limited to that method since it is quite efficient. The goal should be for schools and teachers to fit the method to the child rather than try to fit the child to the method.

Don Deshler and his colleagues at the University of Kansas have developed 'learning strategies' which have proven to be extremely helpful to students with a variety of learning difficulties (Alley & Deshler, 1979; Mercer, 1997). The strategies, taught by teachers who have been trained in their use by Deshler, teach children how to learn. The success of the learning strategies has clearly demonstrated to educators that emphasis on the process of learning needs additional attention.

Expectations. A teacher's expectation greatly influences outcome. If a teacher believes that female students can't be successful in a computer course, then he or she will be more likely to ignore the female students. If a teacher thinks that a homosexual student is not worthy of a good education, that attitude may be communicated to the student and his or her feelings of self worth could be diminished. On the other hand, when teachers believe that all children, regardless of label, can and should succeed, their chances of success will be greatly enhanced.

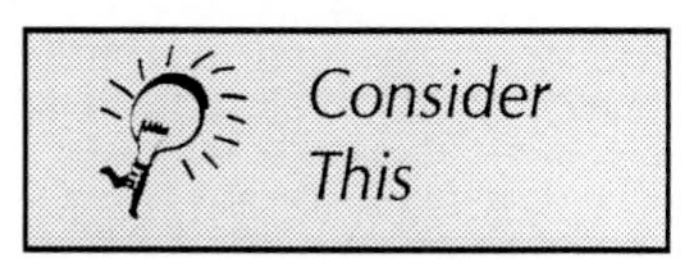

What career options are available for people interesting in promoting human diversity?

Technology. New technology is exciting. For all of the children in our schools who learn differently from most others, whether due to a health disorder, a cultural difference, or a physical or mental disability, today's explosion of technological discoveries is a wonderful asset. Research on the world wide web, computers which change print to the spoken word, computerized translation programs, spelling and grammar checkers, computer access to almost any college or government supported library, all offer new opportunities in education.

Careers. Various career options are available for people interested in promoting fair and equal educational opportunities. Some of these careers include:

- Art Therapy
- Employment Counseling
- Human Resource Counseling
- Mental Health Counseling
- Music Therapy
- Nursing
- Occupational Therapy
- Physical Therapy
- Health Care
- Psychology
- Social Work
- Special Education teaching
- Vocational Rehabilitation counseling
- Anthropology
- Sociology

We urge those of you who have not yet decided upon your professional future to consider employment in a field related to human diversity.

Conclusion

For additional consideration of the many issues regarding diversity in the classroom, the timely and thought provoking works of Schuman and Olufs (1995) and Sleeter and Grant (1994) provide rather thorough discussions and a variety of perspectives. These authors have dealt with many aspects of diversity beyond those discussed here. Schuman and Olufs may be most helpful to those of you who wish to consider the subject of diversity in the college environment. Sleeter and Grant offer many specific suggestions for educational improvements and strategies for successful teaching. Additionally, Schwartz, Conley, and Eaton (1996) have a new text entitled Diverse Learners in the Classroom. This text provides many practical ideas and strategies for those entering the teaching profession. Educating diverse learners is exciting—so read on and keep on learning.

References

Alley, G. & Deshler, D. (1979). *Teaching the learning disabled adolescent: Strategies and methods.* Denver: Love.

Cushner, K., McClelland, A., & Safford, Philip. (1996). *Human diversity in education.* New York: McGraw Hill.

Garcia, E. (1994). *Understanding and meeting the challenge of student cultural diversity.* Boston: Houghton Mifflin.

Grossman, H. (1995). *Teaching in a diverse society.* Needham Heights, MA: Allyn and Bacon.

Kanner, L. (1964). *A history of the care and study of the mentally retarded.* Springfield, IL: Charles C. Thomas.

Landes, A. (Ed.). (1994). *Minorities: A changing role in America.* Wylie, TX: Information Plus.

Mercer, C. (1997). *Students with learning disabilities.* Upper Saddle River, NJ: Merril/Prenctice Hall.

Price, D. (1996). Million-dollar court case sends a clear message to schools: Protect gay students equally or pay. [On-line]. Available: http://www.detnews.com/1996/menu/stories/77022.htm

Schuman, D. & Olufs, D. (1995). *Diversity on campus.* Needham Heights, MA: Allyn and Bacon.

Schwartz, S. (1996). Gay and lesbian students: Issues for special education teachers and teacher educators. Paper presented at The Council for Exceptional Children Conference, Orlando, Florida.

Schwartz, S., Conley, C., & Eaton, L. (1996) *Diverse learners in the classroom.* New York: McGraw Hill.

Sleeter, C. & Grant, C. (1994). *Making choices for multicultural education: Five approaches to race, class, and gender.* New York: Macmillan.

Smith, D. & Luckasson, R. (1992). *Introduction to special education: Teaching in an age of challenge.* Needham Heights, MA: Allyn and Bacon.

Sullivan, A. (1995). *Virtually normal: An argument about homosexuality.* New York: Alfred A. Knopf

Suggested Readings

Driedger, D. (1989). *The last civil rights movement.* New York: St. Martin's Press.

This book traces the history of Disabled People's International.

Cegelka, P. T. & Berdine, W. H. (1995). *Effective instruction for students with learning difficulties.* Boston: Allyn and Bacon.

This provides detailed descriptions for how to program students for success in the educational environment by using effective teaching strategies.

Public Law 94-142, the Education for All Handicapped Children Act of 1975. Public Law 101-476, the Individuals with Disabilities Education Act of 1990.

You can obtain a copy of the laws by writing to your United States senator or congressional representative. The laws are very interesting reading!

Rogers, J. (1994). *Inclusion: Moving beyond our fears.* Bloomington, In: Phi Delta Kappa.

This volume is a compilation of 22 research based articles and professional opinion on one of America's most immediate and least understood educational problems.

Individual Activities

1. Develop a list of those things that are special about special education.

2. Develop a list of opportunities that alternative schooling might allow a student to pursue.

1. Make arrangements so your class can visit and observe special education classes in your local school district. While observing, look for the aspects you listed as being special about special education.

Reactions:

2. Arrange for a panel of teachers with different types of students to visit your class. Ask the panel to discuss some of their experiences with student risk factors.

Reactions:

Reflection Paper 16.1

Reflection Paper 16.1

Assume that you are a political leader and you have to make a decision whether to support or oppose a bill to give additional money to special education. How would you vote and why?

Reflection Paper 16.2

Reflection Paper 16.2

Name three current controversial issues in education relating to diversity and discuss how these issues might be resolved over time.

Notes

Chapter 17 Societal Perspectives

Chapter Sections

- A Vision of Possibilities
- Your Comfort Level
- Community Impact
- Careers with Diverse People
- A Final Self-Check
- The Next Step
- Individual and Group Activities

"How wonderful it is that nobody need wait a single moment before starting to improve the world."
—Anne Frank, diarist

Objectives

- Develop your own strategies for assisting people in dealing with diverse individuals.
- Begin to alter biased attitudes children learn from their parents and from other adults in society
- Establish yourself as a role model for your peers.
- Explore the meaningful contributions you can make to your fellow citizens.

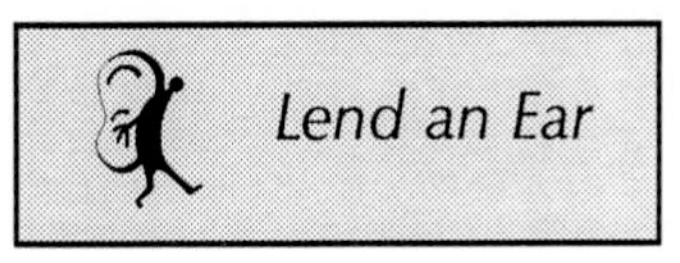

"Every bigot was once a child free from prejudice."
—Sister Mary de Lourdes

A Vision of Possibilities

Although it appears that people have generally become more understanding of human differences over the past several decades, and though we have attempted to institutionalize respect for human difference through legislation, it is still common for individuals to fear and ridicule those who are different. For instance, someone may be critical of a person with a visible disability who eats in a restaurant. Another person may be concerned about a group living home for people who are diverse being established in his or her neighborhood. Prejudice against a co-worker because of race, gender, or sexual orientation is something many of us hear expressed on a regular basis. Myths and negative attitudes about human diversity still exist, and will exist until *we* change them, one person at a time (see Figure 17.1). The easiest person to change is the one you see when you look in the mirror. We hope that this course has been a good beginning for you.

There has been progress toward establishing equal rights for diverse people. However, that does not mean that everyone who interacts with them will be kind, considerate, or comfortable. Part of the problem is that children learn certain attitudes from their parents and from other adults and peers in society. Now you are

Figure 17.1
The only way to make a significant difference in society is to dispel myths, one person at a time.

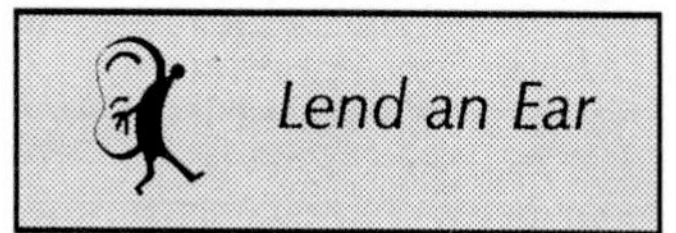

"Tiny ripples of hope... from a million different centers of daring and energy will change the world." —Robert F. Kennedy, political leader

in a position to alter those attitudes for yourself and others. You are equipped with information and can help to insure that society makes additional progress well beyond the laws which have assisted diverse people.

Wouldn't you prefer to live in a society where people don't stare at others who are different? Where everyone, regardless of race, ethnic, or religious background, gender, sexual orientation, age, intellectual ability, nationality, socioeconomic level, size, or shape is treated with respect? Where people are evaluated on the basis of their skills and their personality, not on the basis of the labels which have been imposed upon them by others? When that happens, the society is richer in every way because we all have the opportunity to contribute to the fullest extent of our capacity. You have the ability now to influence others so that fair treatment is afforded to everyone.

This final chapter will help you to develop strategies for assisting others in dealing with diverse individuals. Realize that you are educated now and that you are in the position to be a diversity advocate. By taking this course, you have prepared yourself to be a leader of others, and others will follow your example since they know you have had an opportunity to study and consider these issues. Each of the sections in this chapter offers you ideas and encouragement as you reconsider your own attitudes and begin to change society. It will be a wonderful and exciting experience to grow in understanding as you help others to do the same!

Your Comfort Level

Each chapter has stressed the importance of being comfortable when you interact with people who are different. Comfort logically follows the reduction of fear as you dispel myths, correct misinformation, and develop knowledge and understanding. Some steps toward becoming more comfortable around people who are different are listed below.

Make a firm personal decision to be comfortable when you interact with anyone different from you. Often, merely having the *intention* to be comfortable will help you to actually feel comfortable. You might begin your day with an affirmation such as this: "Today I hope to meet someone different from me so that I can discover an underlying similarity."

Continue to learn more about human diversity. Consider taking additional courses in the many areas of diversity. Classes which offer skills such as sign language will make it possible for you to easily converse with people who use that form of communication and will make you a more expressive person. Showing an interest in the problems of any special group, whether it is a senior citizens' organization, a women's rights group, a local AIDS service organization, or a club for international students, will broaden your perspective and allow you to participate more fully in the world. At your college or in your neighborhood you should be able to find many opportunities for interacting with different types of people.

When you make friends with people who are different, don't be afraid to talk to them about their exeptionality. Ask the questions you want to ask. Raise the issues you want to raise. Openness and honesty is the key to friendship, and it is important to treat people who are different the same way you treat anyone else. Remember that the most important lesson you have learned is that individuals are unique. They are not defined by labels.

Assist different types of people through volunteer organizations. This will offer you another opportunity to get to know people with differences.

How can a person continue to learn more about human diversity?

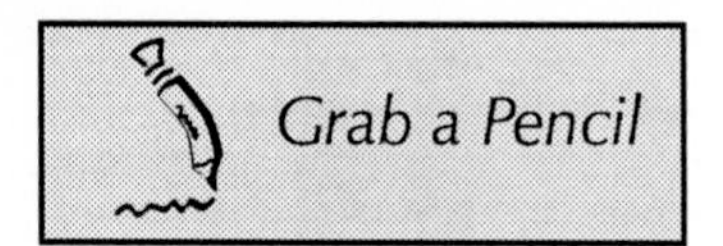

Check out the listing of organizations dealing with human diversity which is included at the end of this textbook. Mark which organizations you would like to contact.

Literacy programs exist in every community, as do programs for economically disadvantaged people, and nearly every other category of exceptionality. Such organizations or agencies will welcome your assistance and will usually have a variety of projects or activities from which you can choose. An annotated listing of organizations dealing with human diversity is included at the end of this book.

Include diverse people among your acquaintances and friends. Don't be afraid to invite someone different to social functions. Remember that differences are not contagious, and even if they were, you and every other person alive is exceptional in some way, in some context. Some of your friends may not be comfortable with your new acquaintance. Their response may be due to their ignorance, and you can use that as an opportunity to help them learn and see from your perspective.

Just be yourself around people who are different. If you are false, they will recognize it. Don't worry about your mood—we all have good and bad days. Don't adjust your vocabulary. Consider the feelings of others as you always do, letting your new knowledge and sensitivity enhance your natural good manners.

Community Impact

Now, or in the future, you may be in the position of having significant influence upon your community. You may belong to civic, religious, and professional organizations. Perhaps you will be a city commissioner, a state legislator, or the President of the United States. You may be a leader in business or industry and have responsibility for hiring or influencing the careers of others. What wonderful opportunities you may have to make a positive impact upon our society.

As a member of a civic, religious, or professional group, you certainly can set an example for others regarding interactions with diverse people. Some individuals with exceptionalities may need your sponsorship to facilitate their membership, or they may need special accommodations for access in order to participate fully.

For instance, if your group is presenting a special speaker and opening the meeting to the public, be sure to hire a sign language interpreter and let that be known in your publicity so that those needing that service will be motivated to attend (see Figure 17.2). Obvious requirements for location such as wheelchair ramps may be overlooked by those unaccustomed to thinking inclusively. You can point out these considerations.

If you become active in politics, you will have many chances to significantly influence the funding of programs, the establishment of procedures for service development and provision, and the design of laws which directly affect diverse people. At the local, state, and federal level, as an active politician or simply as a voter, we encourage you to follow two simple rules. The first is that people who are different deserve every opportunity to lead the normal lives they desire. The second is that they should not be merely tolerated or accepted, but respected as full and important members of the community.

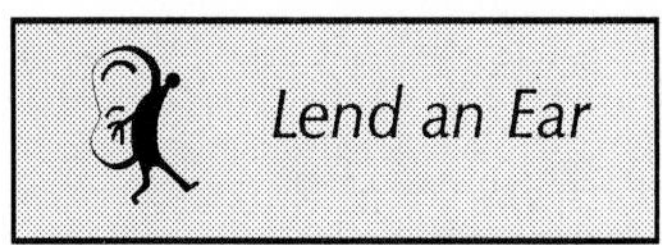

"I will oppose measures that divide, disrespect or diminish our humanity. Our overarching principle should be to promote civility, mutual respect and unity." —Gary Locke, political leader

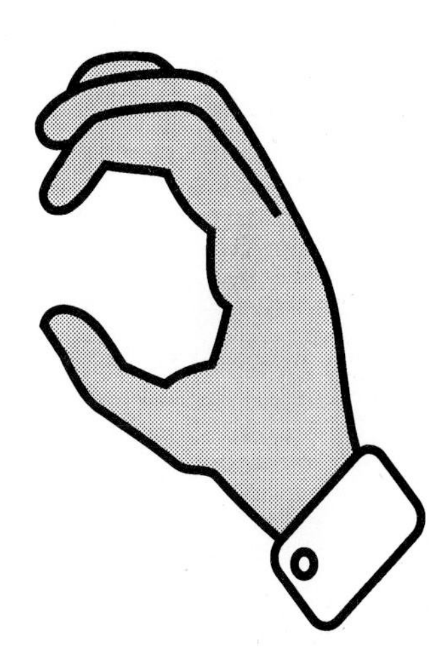

Figure 17.2
See to it that a sign language interpreter is available for all public meetings.

Careers With Diverse People

Before closing your study of human diversity you might consider your future. A possible career path could be in one of the professions dealing with people who are different. Such fields as psychology, sociology, anthropology, political science, counseling, medicine, special education, music therapy, speech and language therapy, occupational or physical

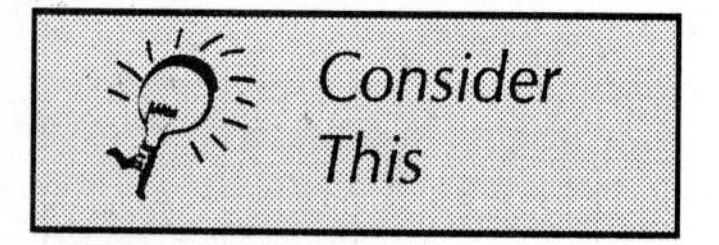

In the exercise on this page, you were asked about which doctor you would want to save your life. What label did you choose? Why?

therapy, sign language interpreting, and many others offer challenging careers which let you make meaningful contributions to your fellow women and men. If you think you might be interested, a good way to begin is to visit some professionals in the area of your choice to explore the possibilities and ask advice on some courses of study.

A Final Self-Check

Now that you are educated and informed about human diversity, here is a test of your new attitudes and ability to respect and assist those who are considered different. Imagine this situation:

> *You are being wheeled into a hospital emergency room. You have sustained a life-threatening injury. Your life depends upon the ability of the lone doctor on duty. Which one of the following labels do you want that doctor to have?*
>
> *Male Female Christian Jewish Muslim*
>
> *Homosexual Heterosexual Bisexual Asian*
>
> *Slim Tall Heavy Short Rich Poor*
>
> *Disabled Caucasion Black Hispanic*
>
> *Hindu Old Young Foreign Expert*

If you value your life, your answer was "Expert." It is doubtful that you would refuse treatment and choose to die if the doctor had any of the other labels coupled with "expert." Since that is true, we know that you, and most everyone else, can ignore labels when it suits our needs. If you can choose to ignore them at all times and in all situations, looking only at the abilities of the person involved, then you have learned to treat individuals with the respect they deserve. When old habits of thinking and fears based on prejudice and ignorance creep in, it is a handy reminder to ask yourself, "If I were in that hospital emergency room and this person were the only doctor on duty, would I want this person to save my life?" Every time you say "yes," you save your own humanity and make the world a much better place.

Reflect upon how you have changed because of your study of human diversity.

The Next Step

Your study of human diversity is at a close. The next step you take is up to you. We hope that you have been inspired to continue discovering more about differences among people. We are confident that you are now equipped to make a difference in the lives of people with whom you come in contact. We are also confident that you can set an example in your community. The sample nondiscrimination policy provided on the right may prove useful in the classrooms, offices, and other organizations to which you belong.

As you strive to learn more about human diversity, we hope that you continue to see the underlying unity among all peoples while at the same time you respect individual uniqueness. The human race has a long way to go before universal understanding becomes the norm, but the only way we'll ever reach that goal is by changing one person at a time, starting with ourselves. If, after this course, you have a better understanding of human diversity and an attitude of respect for persons who are diverse, you have taken a big step toward the goal.

Nondiscrimination Policy

As fellow members of the human race, we support the diversity of all those who make up the stakeholders of our society.

Current nondiscrimination policies and practices of our society already support and celebrate this diversity.

To demonstrate our unity in diversity, we will not accept any kind of prejudicial, obscene, demeaning, abusive, or profane language, gestures, or acts.

Suggested Readings

Gerbner, G. (1990). Communication. *Encyclopedia Americana* (International edition, vol. 7, pp. 423-24). Danbury, CT: Grolier.

This is an insightful overview of the diverse ways in which humans exchange information.

Hirsch, E. D., Kett, J. F., & Trefil, J. (1988). *The dictionary of cultural literacy.* Boston: Houghton Mifflin Co.

This is an essential quick reference for facts and terminology relating to world literature, philosophy, religion, art, history, politics, geography, anthropology, sociology, psychology, science, mathematics, medicine, and technology.

Jampolsky, G. G. (1990). *Love is the answer: Creating positive relationships.* New York: Bantam.

This is a guide to improving interpersonal relationships, love, and peace of mind.

Keyes, K. (1990). *The living love way.* San Francisco: New Dimensions Foundation.

This book discusses the importance of transforming conflict in one's daily life and expressing love wherever and whenever possible. The discussion also provides insights about the possibilities of world peace.

Kronenwetter, M. (1993). *Prejudice in America: Causes and cures.* New York: Franklin Watts.

This book traces the origins of prejudice in America and suggests practical solutions to the far-reaching problem.

Individual Activities

1. What are some things that you have done in the past in regard to diverse people that will be different in your future? List five things.

2. List six things that you have learned during this course that have assisted you in improving your comfort level with human diversity.

1. In a few years all of you are going to hold important positions in your communities. Get together in small groups and discuss what you will do to make your community a better place for *all* people. Present these actions to your entire class.

Reactions:

2. Invite a panel of diverse individuals and parents of diverse people to your class. Discuss and evaluate the suggestions regarding the improvement of your comfort level which were given in the chapter.

Reactions:

Reflection Paper 17.1

Assume that your best friend has a negative attitude toward diverse individuals. What specific guidelines can you practice to assist your friend in improving his or her attitude?

Reflection Paper 17.2

The phrase "your individual comfort level" has come up frequently throughout this textbook. Describe your comfort level at the beginning of this course and compare it to your current comfort level. How have you changed? What prejudices, if any, have you dispelled?

Appendix A: National Diversity Programs and Services

The following resources are categorized according to the chapters of this text. Inclusion here does not necessarily constitute endorsement.

1. General Human Diversity

- Our Diverse World Homepage
http://www.coe.ufl.edu/DiverseWorld/
This is an interactive website where people can learn about every category of human diversity in detail. The goal of this homepage is to unite young people so that together they can preserve, celebrate, and grow from their differences.

- Resisting Defamation
2530 Berryessa Road, No. 616
San Jose, CA 95132
408-995-6545
408-923-5836 fax
This group works toward eliminating any stereotypes, slander, libel, or crimes against persons from different ethnic groups.

- Unity-and-Diversity World Council
5521 Grosvenor Boulevard, Suite 22
Los Angeles, CA 90066-6915
310-577-1968
310-578-1028 fax
This group provides worldwide coordination for cultural, scientific, educational, and religious nonprofit organizations, businesses, and individuals. It fosters "the emergence of a new universal person and civilization based on unity and diversity among all peoples and all life." The Council seeks to aid in establishing a new, worldwide civilization based upon the reality of the whole person by applying the methods and discoveries of modern science coupled with the insights of religion, philosophy, and the arts.

● The Friendship Force
57 Forsythe Street NW, Suite 900
575 S. Tower
Atlanta, GA 30303
404-522-9490
404-688-6148 fax
With 250 regional groups, members in 42 countries promote global understanding through the "force of friendship." A group of citizens flies to a city in another nation to stay in private homes for an exchange period of approximately two weeks, exchanging a cross-section from each community, representative of occupation, race, age, and gender.

● Peace Net
http://www.igc.org/igc/peacenet/index.html
This site explores information and work for positive social change in the areas of peace, social and economic justice, human rights, and the struggle against racism.

● One World
http://www.oneworld.org/
This is a community of over 120 leading global justice organizations under one roof, featuring news, perspectives, discussions, action agendas, and human rights job markets.

● The Council for Exceptional Children
1920 Association Drive
Reston, VA 22091
703-620-3660
800-328-0272
This council advances the quality of education for all exceptional children and improves the conditions under which special educators work. The Council is divided into smaller divisions which address children with behavioral disorders, mental retardation, communication disorders, learning disabilities, physical disabilities, and visual impairments. It also addresses gifted children and culturally and linguistically diverse exceptional learners.

● Worldwide Friendship International
3749 Brice Run Road, Suite A
Randallstown, MD 21133
410-922-2795
410-922-2795 fax
This group bridges gap among people of all nations through correspondence, so that they can learn each other's culture, language, and values. Membership spans 127 countries and encompasses numerous traditions, creeds, colors, ages, and national origins.

2. The Culture of Diversity

● Institute for World Understanding of Peoples, Cultures and Languages
939 Coast Boulevard, 19DE
La Jolla, CA 92037
619-454-0705
The Institute conducts scientific research in establishing methodology for the comparative study of all populations, cultures, and languages. It also studies world organizations, especially in regard to the future of world civilization.

● American Community Cultural Center Association
19 Foothills Drive
Pompton Plains, NJ 07444
201-835-2661
The Association encourages people in all communities to develop cultural centers for the purpose of presenting cultural possibilities for everyone, regardless of economic status or geographic location.

● Federation of American Cultural and Language Communities
666 11th Street NW, Suite 800
Washington, DC 20001
202-387-0600
This is a coalition of ethnic organizations representing Americans of Armenian, French, German, Hispanic, Hungarian, Italian, Japanese, Sicilian, Ukrainian, and Vietnamese descent. They work to address areas of common interest to America's ethnic communities. The group seeks to further the rights of ethnic Americans, especially their cultural and linguistic rights.

3. Diversity and the Law

 American Civil Liberties Union
132 W. 43rd Street
New York, NY 10036
212-944-9800
212-869-9065 fax
With 200 local groups, the ACLU champions the rights set forth in the Bill of Rights of the U.S. Constitution: freedom of speech, press, assembly, and religion; due process of law and fair trial; equality before the law regardless of race, color, sexual orientation, national origin, political opinion, or religious belief.

 Section of Individual Rights and Responsibilities
c/o American Bar Association
1800 M Street NW, South Lobby
Washington, DC 20036
202-331-2280
202-331-2220 fax
The Section concentrates on law and public policy as they relate to civil and constitutional rights, civil liberties, and human rights in the United States and internationally. Its projects include representation of the homeless, people with AIDS, and those facing capital sentences.

4. Racial and Ethnic Diversity

 National Association for Ethnic Studies
Arizona State University
Department of English
PO Box 870302
Tempe, AZ 85287-0302
602-965-2197
602-965-3451 fax
email: naesi@asuvm.inre.asu.edu
This Association promotes research, study, and curriculum design in the field of ethnic studies, especially Native American, Black, Chicano, Puerto Rican, and Asian American.

• International Committee Against Racism
150 W. 28th Street, Room 301
New York, NY 10001
212-255-3959
312-663-9742 fax
Regional Groups: 4. State Groups: 30. Local Groups: 28. The Committee is dedicated to fighting all forms of racism and to building a multiracial society. The group opposes racism in all its economic, social, institutional, and cultural forms and believes racism destroys not only those minorities that are its victims, but all people.

• American Society for Ethnohistory
Department of Anthropology
McGraw Hall, Cornell University
Ithaca, NY 14853
607-277-0109
This Department promotes and encourages original research relating to the cultural history of ethnic groups worldwide.

• National Rainbow Coalition
1700 K Street NW, Suite 800
Washington, D.C. 20006
202-728-1180
http://www.bin.com/assocorg/rainbow/rainbow.htm
This multiracial organization fosters social, racial, and economic justice.

5. Gender and Sexual Orientation

• National Organization for Women
1000 16th Street NW, Suite 700
Washington, DC 20036
202-331-0066
202-331-9002 (TTY)
State Groups: 50. Local Groups: 800. This organization consists of men and women who support "full equality of women in truly equal partnership with men." The group seeks to end prejudice and discrimination against women in government, industry, the professions, churches, political parties, the judiciary, labor unions, education, science, medicine, law, religion, and other fields.

• National Council of Women of the United States
777 United Nations Plaza
New York, NY 10017
212-697-1278
The Council works for the education, participation, and advancement of women in all areas of society. It serves as an information center and clearinghouse for affiliated women's organizations.

• National Organization for Men
11 Park Place
New York, NY 10007
212-686-MALE
212-766-4030
818-791-0578 fax
Regional Groups: 26. Local Groups: 30. The members of this group, men and women, are united in efforts to promote and advance the equal rights of men in matters such as affirmative action programs, alimony, child custody, men's health, child abuse, battered husbands, divorce, educational benefits, military conscription, and veterans' benefits.

• National Congress for Men
4511 Marathon Heights
Adrian, MI 49221-9240
202-FATHERS
State Groups: 50. Local Groups: 80. This is a coalition of organizations and individuals promoting fathers' rights, men's rights, and equality of the sexes. It advocates the validity of traditional male roles in the family and society.

• National Gay and Lesbian Task Force
2320 17th Street NW
Washington, DC 20009
202-332-6483
The Task Force is dedicated to the elimination of prejudice against persons based on their sexual orientation. It assists other associations in working effectively with the homosexual community and engages in direct action for gay freedom and full civil rights.

• Association for the Sexually Harassed
860 Manatawna Avenue
Philadelphia, PA 19128-1113
215-482-3528
This organization sponsors educational programs for school children.

• National Gay Youth Network
PO Box 846
San Francisco, CA 94101-0846
Regional Groups: 74. State Groups: 80. Local Groups: 94. With gay youth support groups, gay student unions, and other interested groups, this group serves as a networking resource for the exchange of information.

• Renaissance Education Association
987 Old Eagle School Road, Suite 719
Wayne, PA 19087
610-975-9119
This association provides support and information about gender issues, including crossdressing, transvestism, transsexualism, and other transgender behavior.

• Equal Rights Advocates
1663 Mission Street, Suite 550
San Francisco, CA 94103
415-621-0672
415-621-6744 fax
This is a public interest law center specializing in sex discrimination cases.

6. Religious Diversity

• National Legal Foundation
6477 College Park Square, Suite 306
Virginia Beach, VA 23464
800-424-4242
804-397-4242
804-420-0855 fax
The Foundation actively litigates in defense of First Amendment liberties, with special focus on religious freedom. It prepares briefs, educational materials, and other publications on church-state issues for lawyers, teachers, and interested individuals.

• National Interreligious Task Force
4724 Cedar Avenue
Philadelphia, PA 19143
215-729-4084
With 20 local groups, this organization sponsors workshops on human rights and religious liberty.

• Religion in American Life
2 Queenston Place, Room 200
Princeton, NJ 08540
800-428-8292
609-921-3639
609-921-0551 fax
Leaders from Protestant, Roman Catholic, Eastern Orthodox, Unitarian, Muslim, Jewish, and other groups sponsor programs that support the contributions religion makes in American life.

• Coalition for Religious Freedom
5817 Dawes Avenue
Alexandria, VA 22311-1114
This interreligious organization seeks to preserve First Amendment rights to protect free exercise of religion.

• Rockford Institute Center on Religion and Society
2275 Half Day Road, Suite 350
Deerfield, IL 60015
800-383-0680
708-317-8062
708-317-8141 fax
This interreligious research and educational organization focuses on issues of culture and religion in society.

• Facets of Religion
http://sunfly.ub.uni-freiburg.de/religion/
This colorful and informative Internet homepage discusses all major religions of the world in depth, providing links to other sites of interest.

7. Socioeconomic Diversity

🍎 Coalition for Economic Survival
1296 N. Fairfax Avenue
Los Angeles, CA 90046
213-656-4410
The Coalition addresses the economic concerns of senior citizens and low-income families, especially issues dealing with rent control, tenants' rights, and affordable housing.

🍎 Center for Community Change
1000 Wisconsin Avenue, NW
Washington, DC 20007
202-342-0519
415-982-0346 (San Francisco office)
202-342-1132 fax
The center assists community groups of urban and rural poor in making positive changes in their communities. It focuses attention on national issues dealing with human poverty and works to make government more responsive to the needs of the poor.

🍎 National Center for Children in Poverty
Columbia University School of Public Health
154 Haven Avenue
New York, NY 10032
212-927-8793
212-927-9162 fax
This organization works to strengthen programs that improve the health and development of children living in poverty.

8. Physical Diversity

🍎 Council on Size and Weight Discrimination
PO Box 305
Mount Marion, NY 12456
914-679-1209
The goal of this council is to influence public policy and opinion in order to end oppression based on discriminatory standards of body weight, size, or shape.

• National Association to Advance Fat Acceptance
PO Box 188620
Sacramento, CA 95818
800-442-1214
916-558-6880
916-558-6881 fax
Local chapters: 50. This association is dedicated to improving the quality of life for people who are fat by working to eliminate discrimination based on body size and provide people who are fat with the tools for self-empowerment through public education, advocacy, and member support. The group disseminates information about the sociological, psychological, legal, medical, and physiological aspects of being fat.

• Little People of America
7238 Piedmont Drive
Dallas, TX 75227-9324
800-24-DWARF
214-388-9576
Regional groups: 12. Local groups: 50. This group provides fellowship, interchange of ideas, moral support, and solutions to unique problems of little people. It aids in the exchange of information on medical treatment, employment, clothing, shoes, and education.

• National Information Center for Children and Youth with Disabilities
PO Box 1492
Washington, DC 22013
202-884-8200 (voice, TDD)
800-695-0285 (voice, TDD)
This center collects and shares information and ideas that are helpful to children and youth with disabilities and to people who care for and about them.

• National Association of the Physically Handicapped, Inc.
Bethesda Scarlet Oaks, No. GA4
440 Lafayette Avenue
Cincinnati, OH 45220-1000
513-961-8040
This association advances the social, economic, and physical welfare of persons who are physically handicapped in the U.S. and develops awareness of the needs of people who are physically disabled and supports legislation for their benefit.

• Administration on Developmental Disabilities
Office of Human Development Services
U.S. Department of Health and Human Services
200 Independence Avenue, SW
Washington, DC 20201
202-690-5504
202-245-2890
This office administers the Developmental Disabilities Assistance and Bill of Rights Act, whose programs and services assist persons with developmental disabilities to achieve independence, productivity, and integration into the community.

• National Association of Child Advocates
1625 K Street NW, Suite 510
Washington, DC 20006
202-828-6950
202-828-6956 fax
State organizations: 40. This association works for the safety, security, health and education for all America's children by building and supporting state and community-based independent child advocacy organizations.

• Federation for Children with Special Needs
95 Berkeley Street, Suite 014
Boston, MA 02116
617-482-2915
413-562-5521
617-695-2939 fax
This is a coalition of eleven statewide parent organizations that act on behalf of children and adults with a variety of special needs.

• National Alliance of Senior Citizens
1700 18th Street NW, Suite 401
Washington, DC 20009
202-986-0117
202-986-2974 fax
The Alliance advocates the advancement of senior Americans. It seeks to inform the American public of the needs of senior citizens and of the programs and policies being carried out by the government.

 Paralyzed Veterans of America
801 18th Street NW
Washington, DC 20006
202-872-1300
This organization has excellent materials for persons who are wheelchair users.

 United Nations Children's Fund
3 United Nations Plz.
New York, NY 10017
1-800-FOR-KIDS
212-326-7000
UNICEF works for sustainable human development to ensure the survival, protection, and development of children around the world.

 National Easter Seal Society
230 West Monroe
Chicago, IL 60606
800-221-6827
312-726-1494 fax
This organization has excellent materials regarding architectural barriers and transportation of persons with disabilities and sponsors camps for family respite care.

9. Learning Differences

 Learning Disabilities Association of America
4156 Library Road
Pittsburgh, PA 15234
412-341-1515
412-344-0224 fax
State chapters: 50. Local chapters: 750. This is a national information center and referral services with hundreds of affiliates. It offers a free information packet and publishes a newsletter and journal.

 National Center for Learning Disabilities
381 Park Avenue South, Suite 1420
New York, NY 10016
212-545-7510
This center provides financial support for research on learning disabilities.

• Council for Learning Disabilities
PO Box 40303
Overland Park, KS 66204
913-492-8755
913-492-2546 fax
This council works with individuals who have learning disabilities and aids all LD educators in the exchange of information through publications and conferences.

10. Intellectual Differences

• National Association for Gifted Children
1707 L Street NW 550
Washington, DC 20036
202-785-4268
State Groups: 50. This association advances interest in programs for children who are gifted. It seeks to further educate those who are gifted and to enhance their potential creativity and Distributes information to teachers and parents on the development of the gifted child.

• American Association on Mental Retardation
444 North Capitol Street NW, Suite 846
Washington, DC 20001-1512
202-387-1968
800-424-3688
202-387-2193 fax
This is an interdisciplinary association of professionals and individuals concerned about the field of mental retardation. It promotes the well-being of individuals with mental retardation and supports those who work in the field.

• American Mensa Limited
201 Main Street, Suite 1101
Ft. Worth, TX 76102
800-66MENSA
718-934-3700
This is a society for individuals who have established by score in a standard intelligence test that their intelligence is higher than that of 98% of the population.

• The Arc
http://TheArc.org
This Internet homepage provides a wealth of information on developmental delays and mental retardation, with numerous links to other areas of interest.

• National Association for Creative Children and Adults
8080 Springvalley Drive
Cincinnati, OH 45236
513-631-1777
This association is a group of people that meet to share creative ideas and activities.

11. Health Diversity

• National Center for Health Education
72 Spring Street, Suite 208
New York, NY 10012-4019
212-334-9845 fax
This organization promotes health education in schools and manages a comprehensive school health education curriculum called "Growing Healthy."

• ODPHP National Health Information Center
PO Box 1133
Washington, DC 20013-1133
800-336-4797
301-565-4167
301-984-4256 fax
This center aids consumers in locating health information. The group is funded by the Office of Disease Prevention and Health Promotion, Public Health Service, Department of Health and Human Services.

• CDC National AIDS Clearinghouse
PO Box 6003
Rockville, MD 20849-6003
800-458-5231
Operated by the Center for Disease Control, this hotline provides current information about AIDS and HIV.

● National Organization for Rare Disorders
100 Rt. 37
PO Box 8923
New Fairfield, CT 06812
203-746-6518
203-746-6481 fax
203-746-6927 (TDD)
This is a clearinghouse for information on over 5,000 little-known disorders affecting some 20 million Americans. It is committed to the identification, treatment, and cure of rare disorders through programs of education, advocacy, research, and service.

● Association for the Care of Children's Health
7910 Woodmont Avenue, Suite 300
Bethesda, MD 20814
301-654-6549
301-986-4553 fax
This association advocates family-centered, psychosocially sound, and developmentally appropriate health care for children. It promotes meaningful collaboration among families and professionals across all disciplines to plan, coordinate, deliver, and evaluate children's health care systems.

● American Cancer Society
1599 Clifton Road NE
Atlanta, GA 30329
800-ACS-2345
404-320-3333
This is a nationwide voluntary health organization dedicated to eliminating cancer as a major health problem and preventing cancer, saving the lives of persons with cancer, and diminishing suffering through research, education, and service.

● Epilepsy Foundation of America
4351 Garden City Drive
Landover, MD 20785
301-459-3700
800-332-1000 (information)
800-332-4050 (library)
301-577-2684 fax
This foundation is dedicated to the well-being of persons with epilepsy. It sponsors research and provides toll-free information to lay and professional inquirers. Local affiliates offer support groups, employment assistance, and other direct services.

• National Center for Chronic Disease Prevention and Health Promotion
http://www.social.com/health/nhic/data/hr000/hr0069.html
This Internet page provides information on prevention chronic diseases.

• Muscular Dystrophy Association
3300 East Sunrise Drive
Tucson, AZ 85718
520-529-2000
This association provides comprehensive patient care throughout its nationwide network of 230 MDA clinics. It also supports an international research program to find the causes and treatments for muscular dystrophy and related neuromuscular diseases.

• American Lung Association
1740 Broadway
New York, NY 10019
212-315-8700
800-LUNG-USA
This association is dedicated to the conquest of lung disease and the promotion of lung health.

• Immune Deficiency Foundation
25 W. Chesapeake Avenue, Suite 206
Towson, MD 21204
410-321-6647
800-296-4499
410-321-9165 fax
This foundation supports research and education for the primary immune deficiency diseases and offers various publications on these diseases.

• Juvenile Diabetes Foundation International
120 Wall Street
New York, NY 10015-3904
212-889-7575
800-JDF-CURE
212-725-7259 fax
This is a voluntary health agency founded by parents of diabetic children who were convinced that, through research, diabetes could be cured.

• National Multiple Sclerosis Society
733 3rd Ave., Sixth Floor
New York, NY 10017
212-986-3240
212-986-7981 fax
This group funds research into causes and cures for MS, provides a variety of publications, and has local chapters throughout the country.

12. Communication Diversity

• Voice Foundation
1721 Pine Street
Philadelphia, PA 19103
215-735-7999
215-735-9293 fax
This foundation supports programs of professional education, scientific research, and public information essential to solving vocal problems.

• American Speech-Language-Hearing Association
10801 Rockville Pike
Rockville, MD 20852
800-638-8255
301-897-5700 (Voice or TDD)
This association provides information and referral on a broad range of speech, language, and hearing disorders.

• National Center for Stuttering
200 East 33rd Street
New York, NY 10016
800-221-2483
212-532-1460
This center provides information to parents of children who stutter, offers treatment for children and adults who stutter, and provides training on the latest practices and theories for speech pathologists.

• Orton Dyslexia Society, Inc.
Chester Building, Suite 382
8600 LaSalle Road
Baltimore, MD 21286-2044
800-ABCD123
410-296-0232
410-321-5069 fax
This organization honors Samuel T. Orton, a physician who studied children with language disorders. It combines the interests of both medical and educational professionals interested in dyslexia and language-learning disorders. It sponsors free referrals for diagnosis and treatment. It also seeks to educate the public about dyslexia and support efforts to enhance the self-worth of persons with dyslexia.

13. Behavior and Personality Diversity

• Autism Society of America
7910 Woodmont Avenue, Suite 650
Bethesda, MD 20814-3015
800-3-AUTISM
301-657-0881
301-657-0869 fax
Local chapters: 160. This is a national umbrella organization serving the needs of autistic citizens of all ages. It provides information and referrals to parents, professionals, and individuals.

• Children and Adults with Attention Deficit Disorders
499 NW 70th Avenue, Suite 101
Plantation, FL 33317
305-587-3700
305-587-4599 fax
This is a national alliance of 625 parent organizations which provides information to parents of children with attention deficit disorders.

• Internet Psychology Resources
http://www.gasov.edu/psychweb/resource/bytopic.htm
This Internet homepage presents many topics on behavior, including autism, ADD, depression, and personality.

14. Sensory Diversity

● Alexander Graham Bell Association for the Deaf
3417 Volta Place NW
Washington, DC 20007
202-337-5220
This association of 17 state and 110 local groups encourages people with hearing impairments to communicate by developing maximal use of residual hearing, speech-writing, and speech and language skills. It also promotes better public understanding of hearing loss in children and adults, and helps oral deaf adults and parents of hearing impaired children.

● National Association of the Deaf
814 Thayer Avenue
Silver Spring, MD 20910-4500
301-587-1788 (voice)
301-587-1789 (TDD)
This is the largest consumer organization of disabled persons in the United States, with more than 22,000 members and 51 affiliated state associations. It serves as an advocate for the millions of deaf and hard-of-hearing people in America.

● American Council of the Blind
1155 15th Street NW, Suite 720
Washington, DC 20005
202-467-5081
The Council advocates legislation for persons who are blind. Priority areas of advocacy include civil rights, social security and supplemental income, national health insurance, rehabilitation, eye research, and technology.

● Association for the Education and Rehabilitation of the Blind and Visually Impaired
206 N. Washington Street, Suite 320
Alexandria, VA 22314
703-548-1884
Regional Groups: 7. State Groups: 44. This is the only professional membership organization dedicated to the advancement of education and rehabilitation of children and adults who are blind and visually impaired.

● Deaf-Blind Online
http://198.234.201.48/dbonline.html
This Internet homepage provides a list of pointers to informative homepages dealing with visual and hearing impairments.

15. Family Perspectives

● Institute of Marriage and Family Relations
6116 Rolling Road, Suite 306
Springfield, VA 22152
703-569-2400
703-569-7248 fax
Offering professionally staffed diagnostic, treatment, counseling, and education centers, this institute endeavors to assist individuals and families in coping with and working through problems in family life and relationships.

● Family Resource Coalition
200 South Michigan Avenue, Suite 1600
Chicago, IL 60604
312-341-0900
312-341-9361 fax
This is a national coalition of community-based family support organizations concerned with parenting, family issues, child development, education, and community life.

● National Council on Family Relations
3989 Central Avenue NE, Suite 550
Minneapolis, MN 55421
612-781-9331
612-781-9348 fax
Regional Groups: 3. State Groups: 41. This is a group of family life professionals, including clergy, counselors, educators, home economists, lawyers, nurses, librarians, physicians, psychologists, social workers, sociologists, and researchers. It seeks to advance marriage and family life through consultation, conferences, and the dissemination of information and research. The group specializes in family health issues, ethnic minorities, and religion and family life.

16. Educational Perspectives

American Federation of Teachers
555 New Jersey Avenue NW
Washington, DC 20001
800-238-1133
202-879-4400
202-879-4556 fax
Local Groups: 2,200. This federation assists teachers, educational organizations, and community organizations to work effectively with children.

National Education Association
1201 16th Street NW
Washington, DC 20036
202-833-4000
202-822-7621 fax
State Groups: 53. Local Groups: 12,000. This is a professional organization and union of elementary, secondary, college and university teachers, as well as administrators, principals, counselors, and others concerned with education. Special committees address civil rights, minority affairs, international relations, and women's concerns.

17. Human Diversity in Society

People to People International
501 E. Armour Boulevard
Kansas City, MO 64109
816-531-4701
816-561-7502 fax
http://www.ptp.org/
This is a multinational, nongovernmental, nonpolitical corporation of individuals communicating with each other through personal contact, letters, and travel. The group promotes international friendship and understanding.

Notes

Appendix B:
Sign Language

Artwork by Robert Stiff, B.S.

Sign languages are complex communication systems that incorporate hand movement combinations, facial espressions, and body language to express whole words, thoughts, and concepts rather than spell them out one letter at a time. Sign languages are not merely systems of gestures. They are true languages with their own complicated grammar and rules of handshape, hand location, movement, facial espression, and body language.

Sign language can be both highly concise and highly poetic. With just one handshape—the thumb and little finger stretched out and the first finger pointing forward—a person can make an airplane take off, experience turbulence and engine trouble, circle an airport, and come in for a bumpy landing. One can sign that entire sentence in a fraction of the time it would take to say it aloud (Walker, 1986).

There is no universal sign language. Sign languages differ from one another just as spoken languages differ around the world. American Sign Language (ASL) is one of the most commonly used, featuring approximately 6,000 signs. ASL is popular because it has a long history of use and is easy to master. Though this sign language is called "American," it is not English and is in fact more similar to Chinese because its signs represent concepts rather than single words (Hardman, Drew, Egan, & Wolf, 1993).

As opposed to sign languages, sign systems use manual gestures which attempt to create visual equivalents of spoken language. Finger spelling is a sign system that uses all 26 letters of the alphabet to spell out individual words. Finger spelling frequently supplements sign languages. For example, when there is no sign for a particular word using ASL, a person may resort to finger spelling. A number of sign systems are used in the United States: Seeing Exact English, Linguistics of Visual English, and Signed Exact English.

Knowing sign language will enable you to communicate with people with hearing impairments who use sign language as their means of communication. Another important reason to learn sign language is that it will enrich all your other modes of expression. Just as learning another language increases your vocabulary and helps you to understand the way other people think and communicate, so too will learning sign language enrich your life.

As an introduction to signing, you will find drawings of finger spelling on the following pages. Practice these with a friend or in a mirror. While practicing, say the word (not the letters) as you finger spell. Practice the letters in random order so that you master them individually rather than depend upon the sequence.

After you learn finger spelling, try your hand at learning some common sign language words which are presented in this chapter. Then practice these words with a friend.

References

Hardman, M., Drew, C., Egan, M., & Wold, B. (1993). *Human exceptionality: Society, school, and family.* Boston: Allyn and Bacon.

Walker, L. A. (1986). *A loss for words: The story of deafness in a family.* New York: Harper and Row.

Suggested Readings

Bornstein, H., Saulnier, K., & Hamilton, L. (1983). *The comprehensive signed English dictionary.* Washington, D.C.: Kendall Green.

Costello, E. (1983). *Signing: How to speak with your hands.* New York: Bantam.

Riekehof, L. (1987). *The joy of signing.* Springfield, MO: Gospel.

Sternberg, M. (1990). *American sign language: A concise dictionary.* New York: Harper and Row.

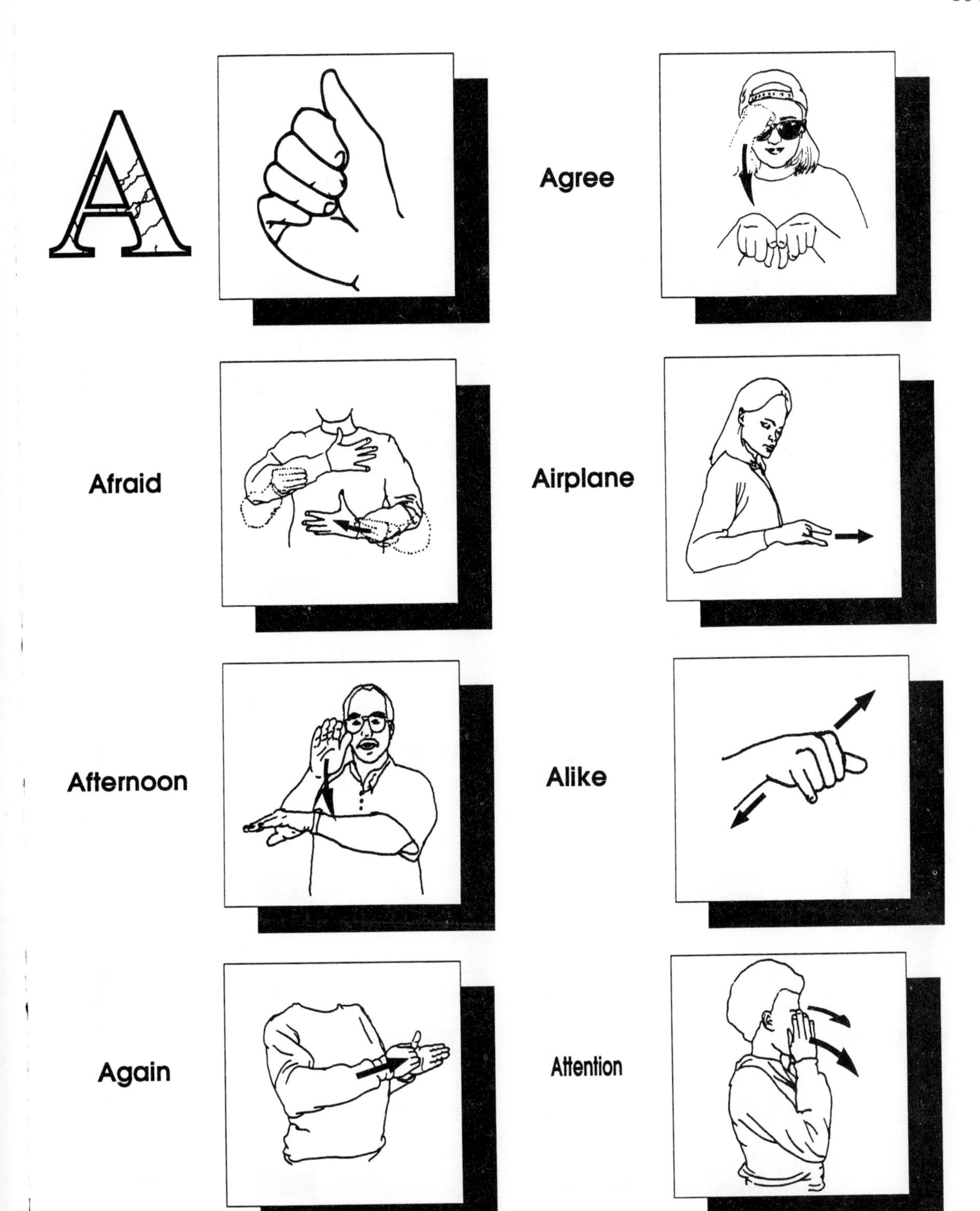
A
Agree
Afraid
Airplane
Afternoon
Alike
Again
Attention

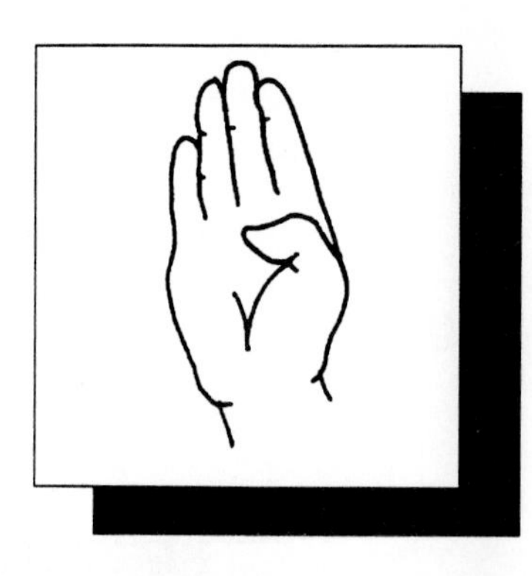

Beautiful

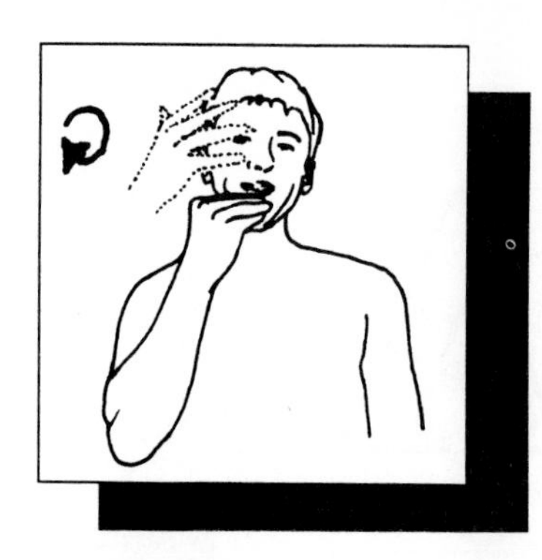

Baby

Beer

Bad

Bicycle

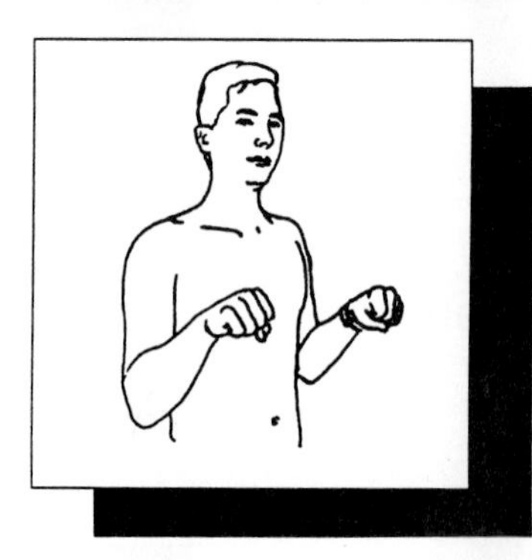

Bath Room

Blind

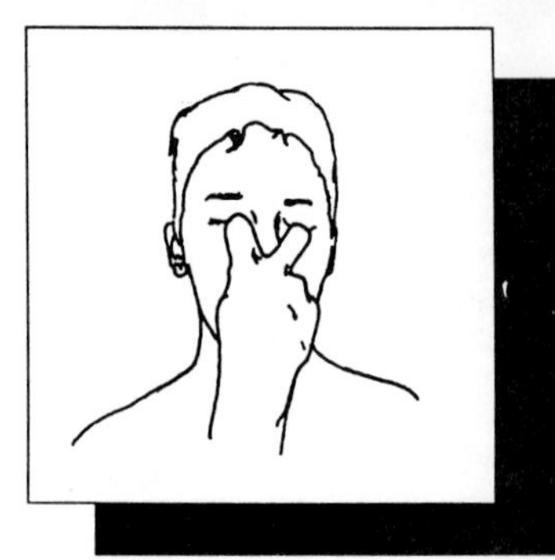

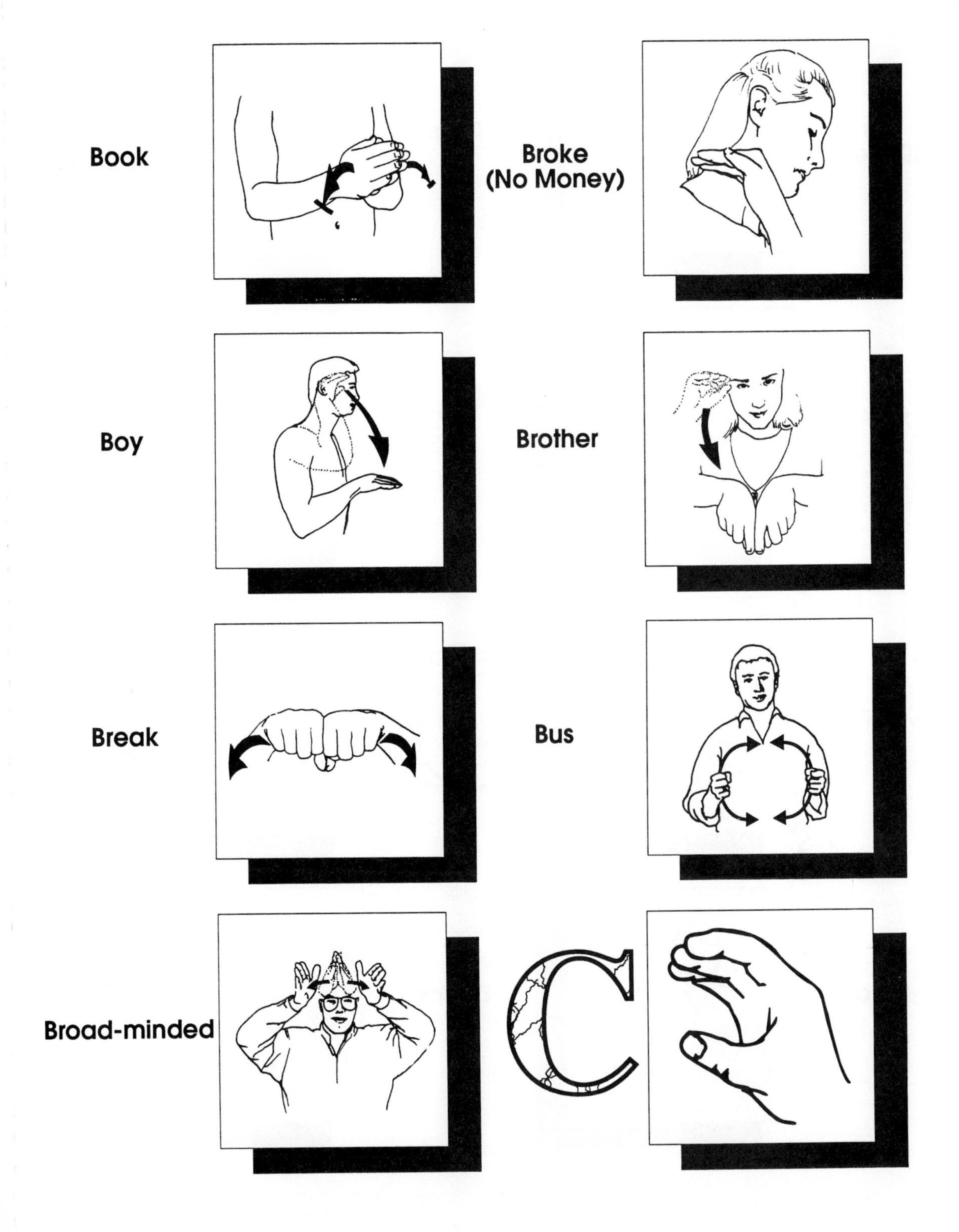
Book
Broke
(No Money)
Boy
Brother
Break
Bus
Broad-minded
C

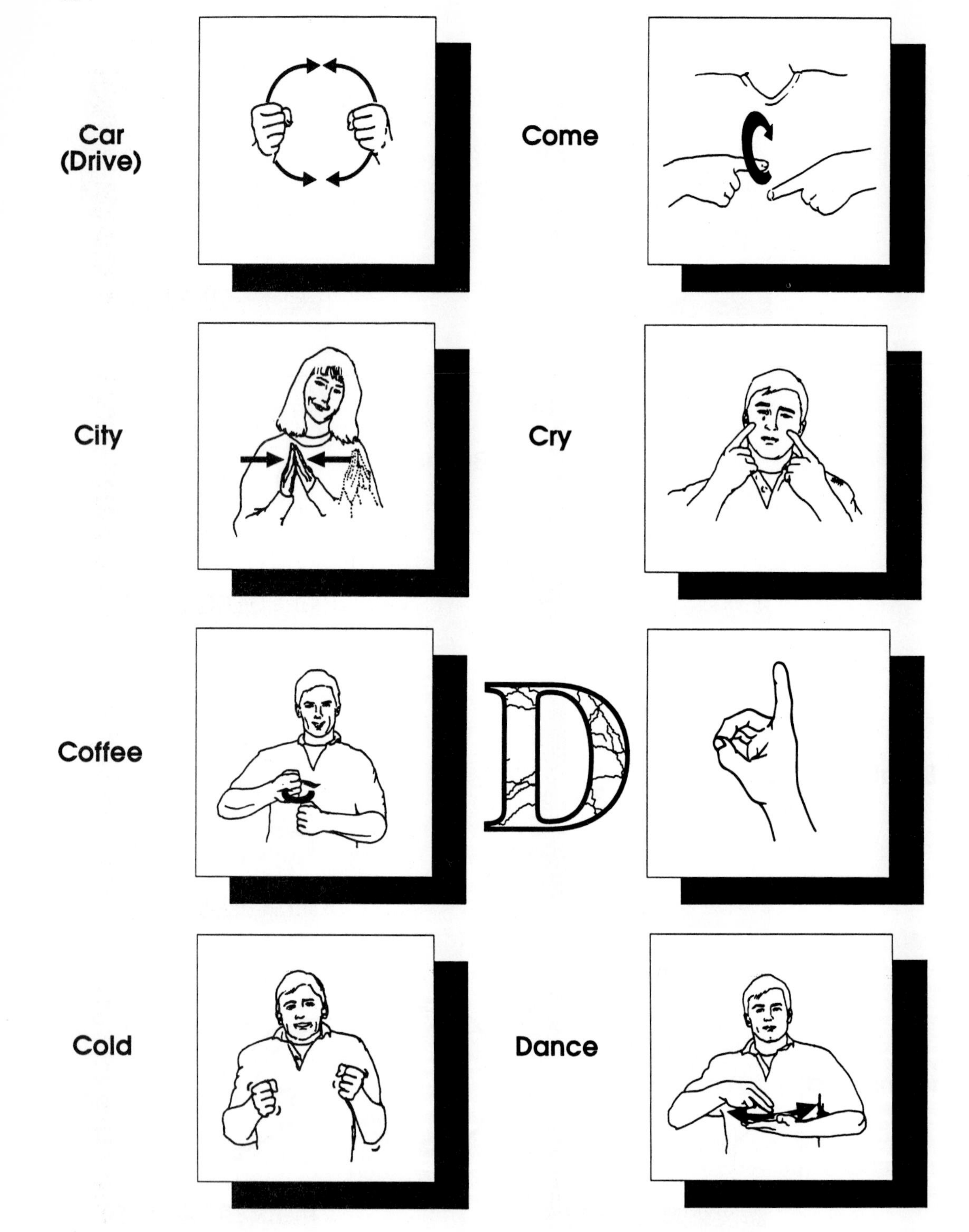
Car (Drive)
Come
City
Cry
Coffee
D
Cold
Dance

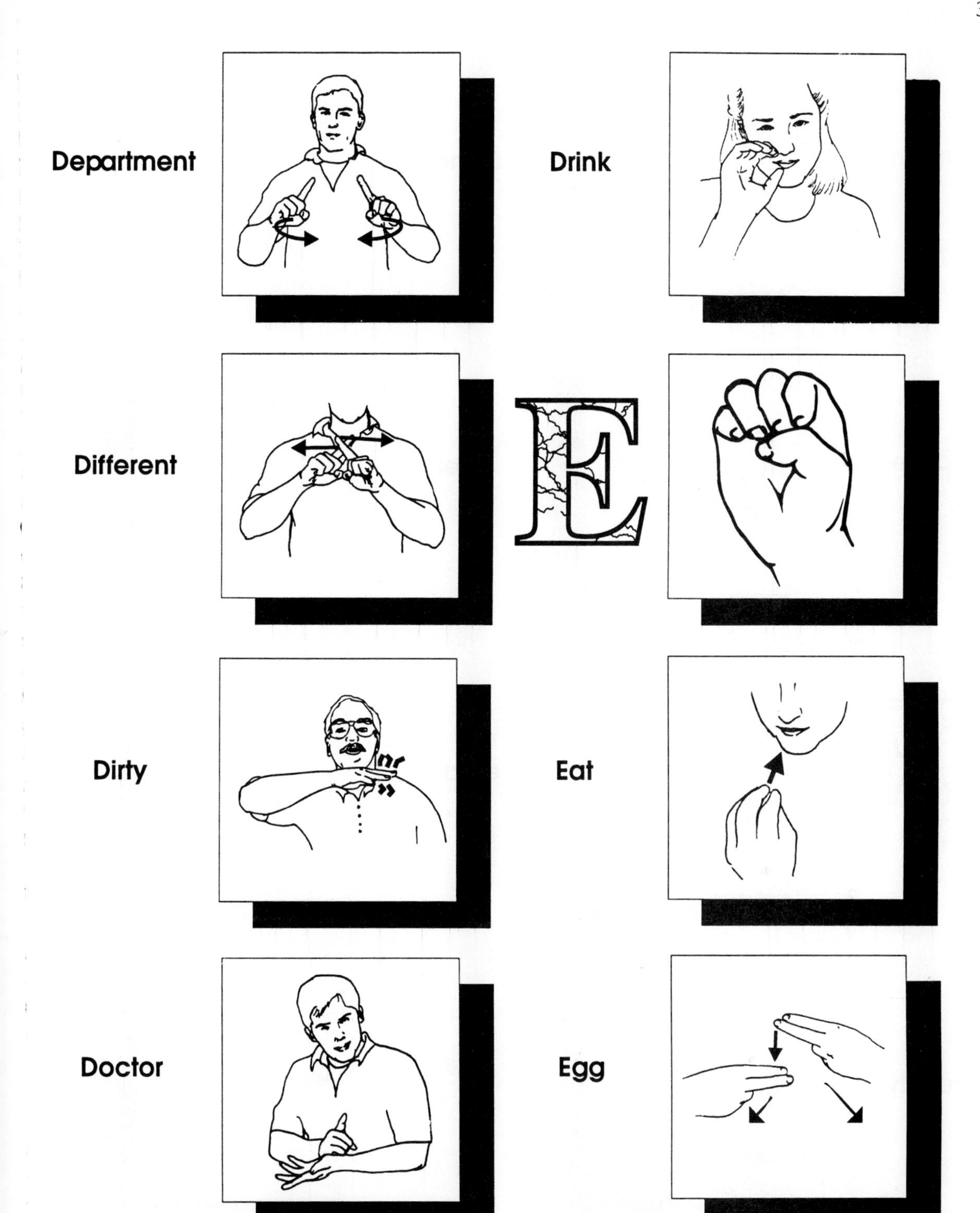
Department
Drink
Different
E
Dirty
Eat
Doctor
Egg

Embarrass

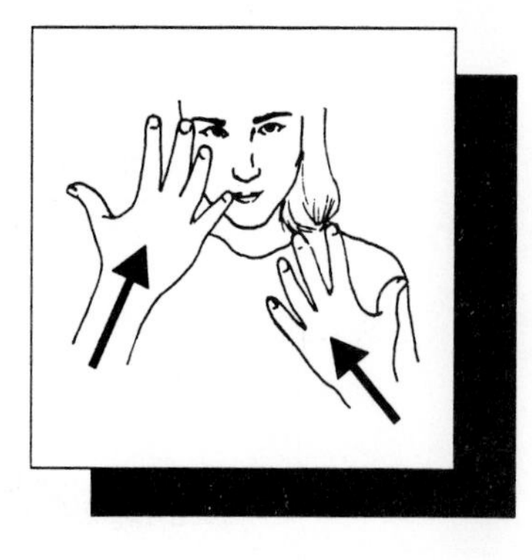

Father

Experience

Feel

F

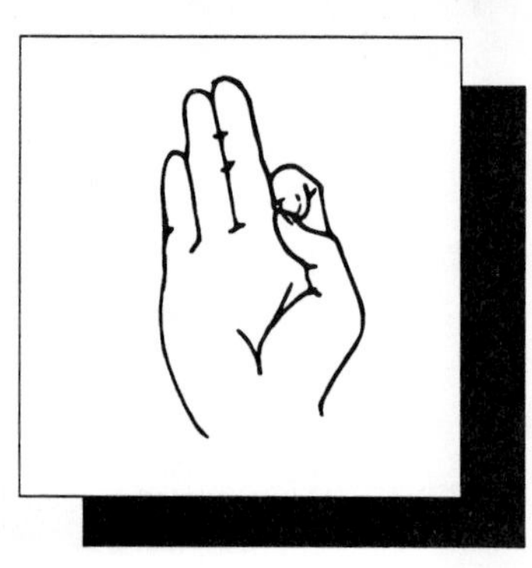

Female

Family

Fine

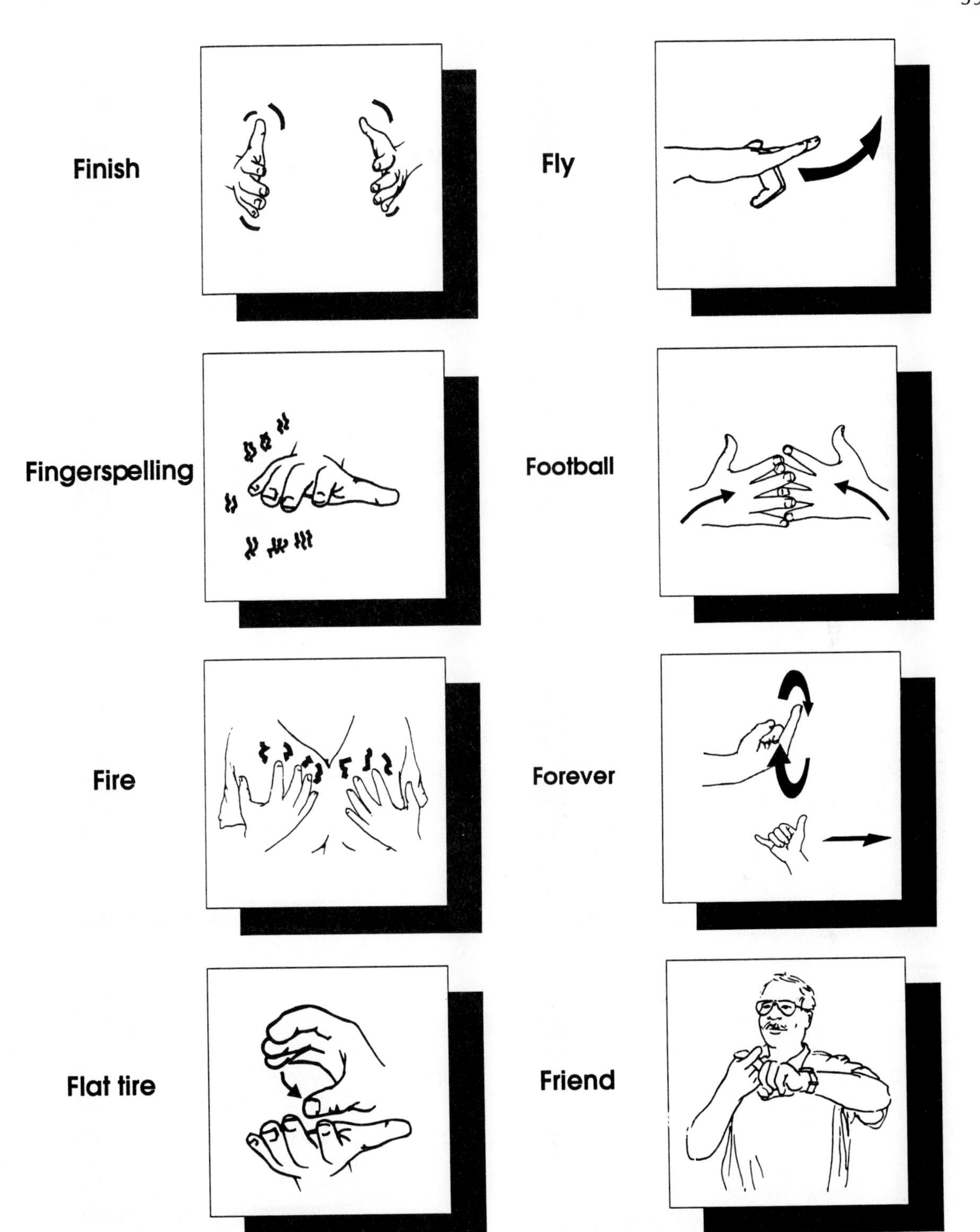
Finish
Fly
Fingerspelling
Football
Fire
Forever
Flat tire
Friend

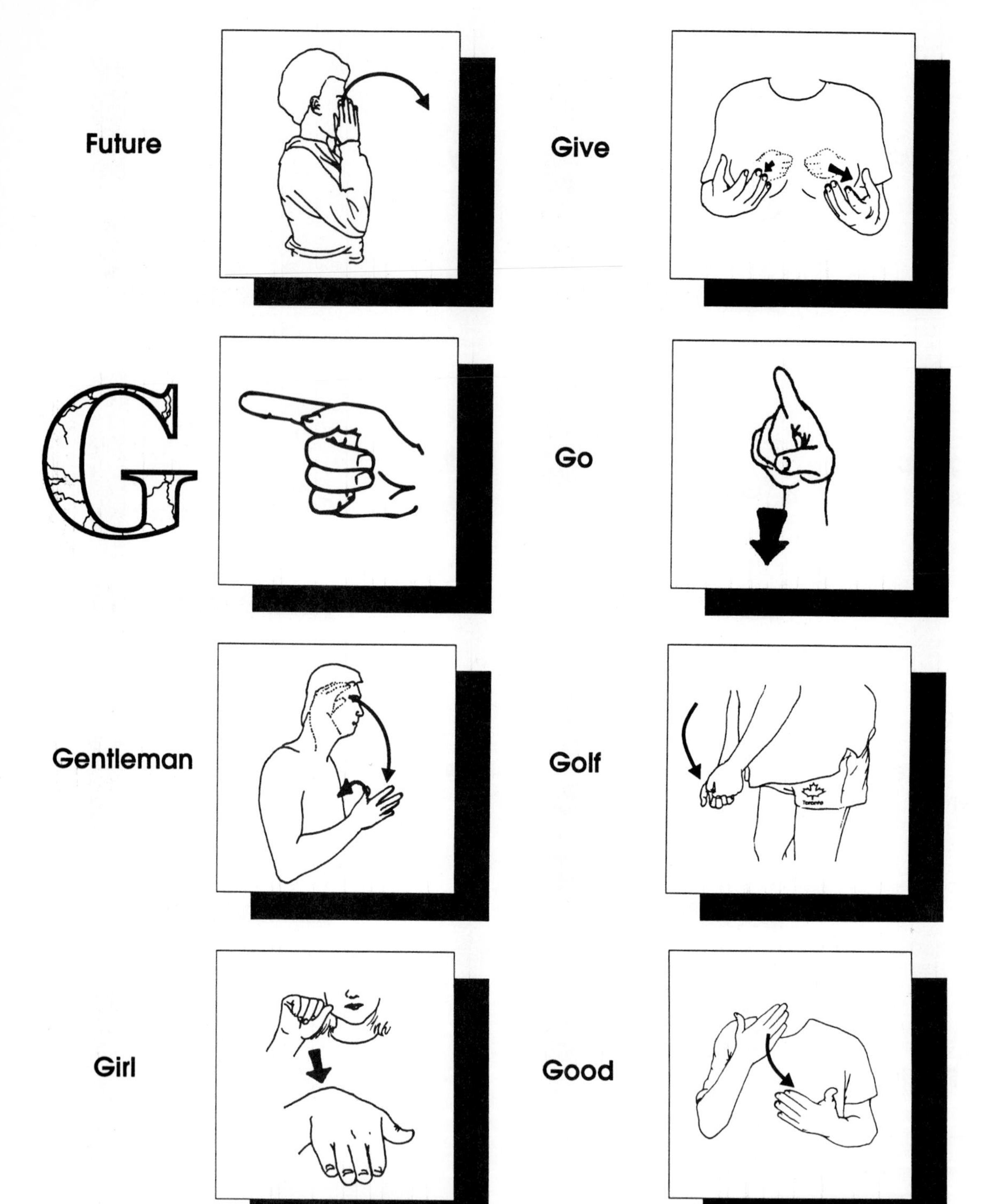
Future
Give
G
Go
Gentleman
Golf
Girl
Good

Grandfather

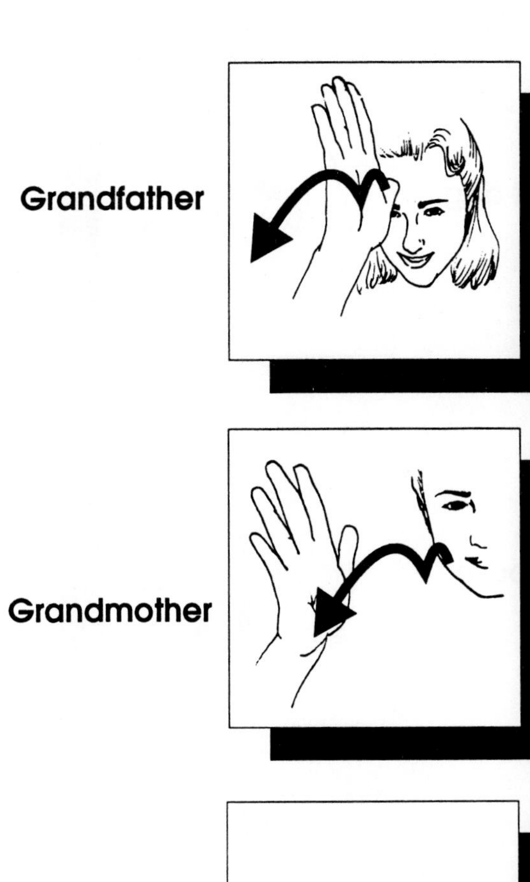

Grandmother

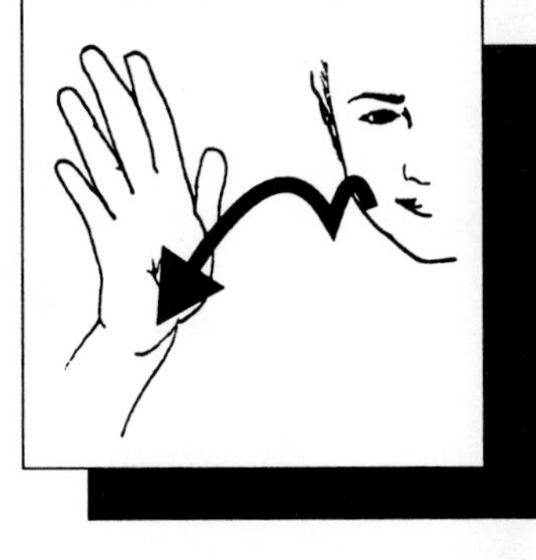

Group

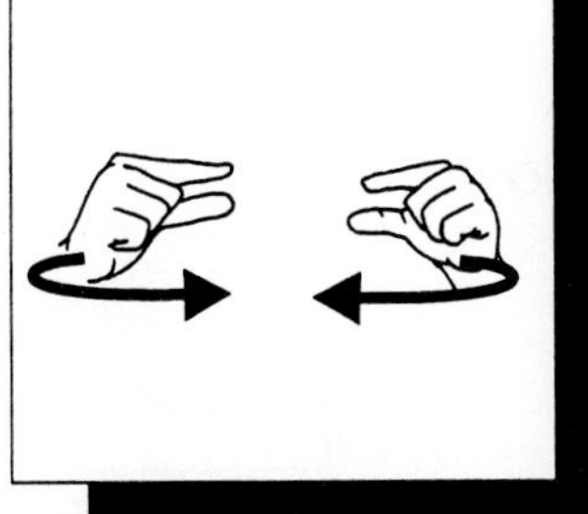

H

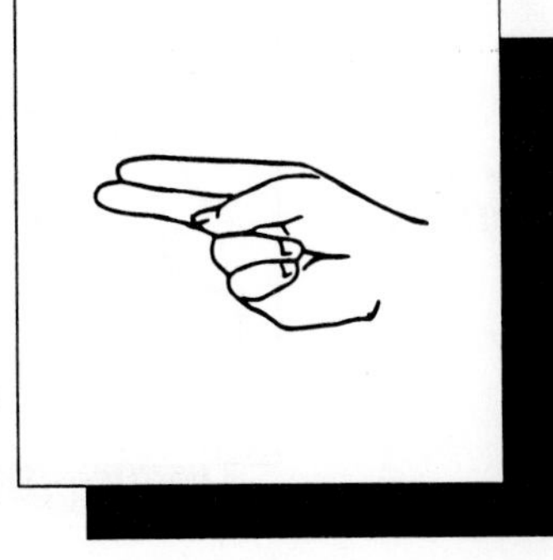

Hamburger

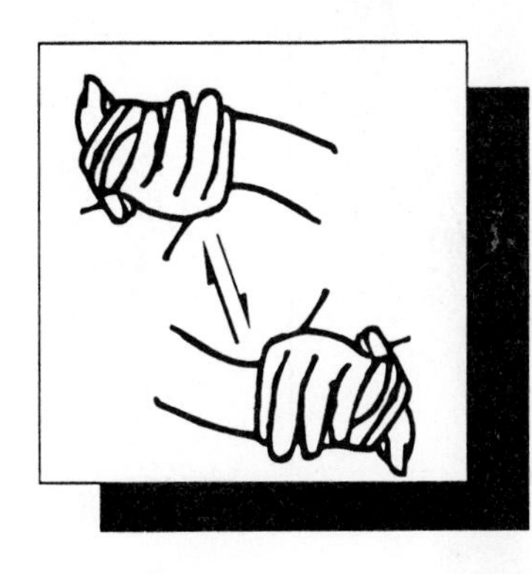

Happy

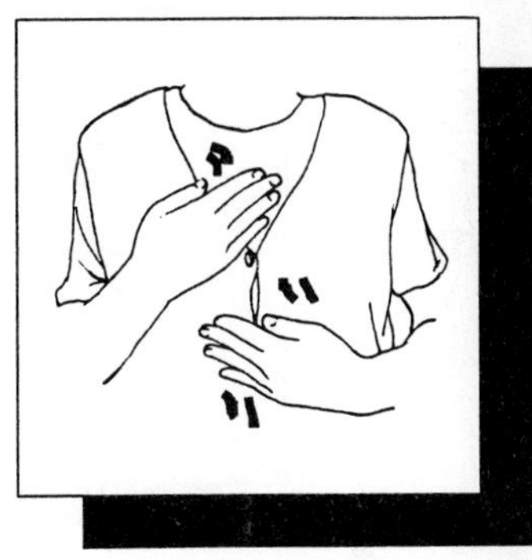

Help

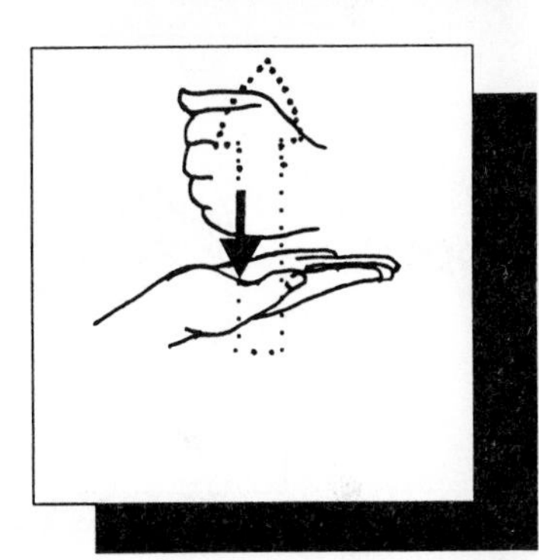

Her

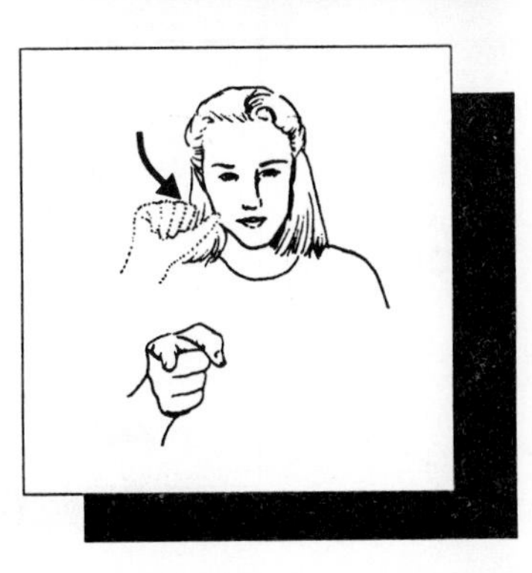

Him

Hot dog

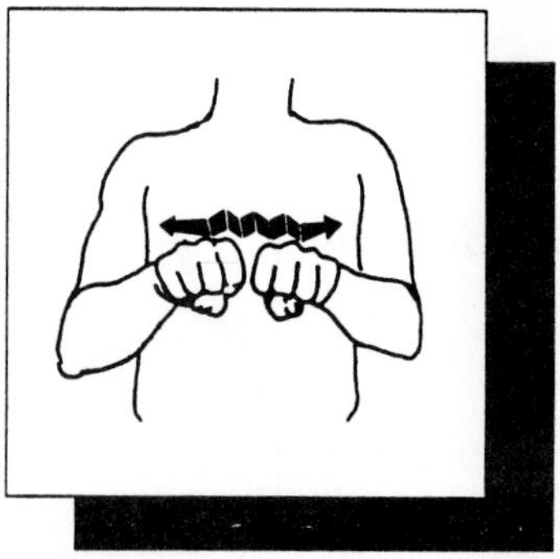

Home

House

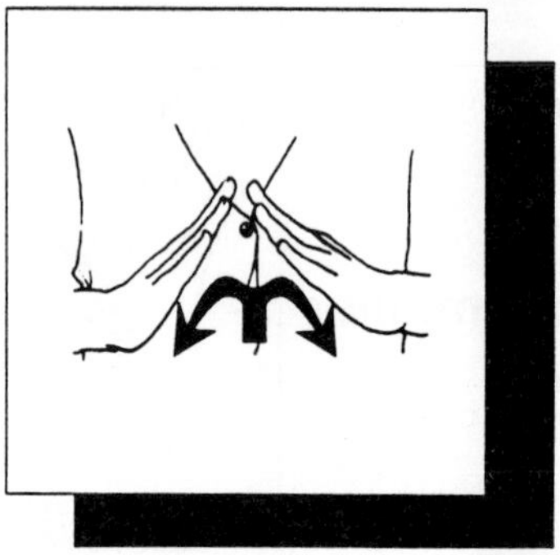

Hope

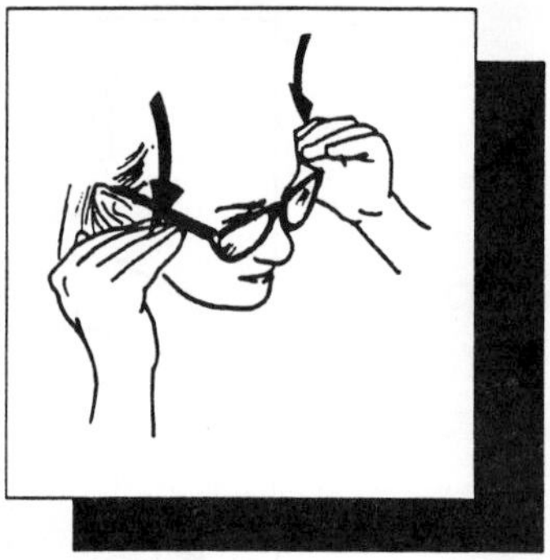

How

Hot

Hungry

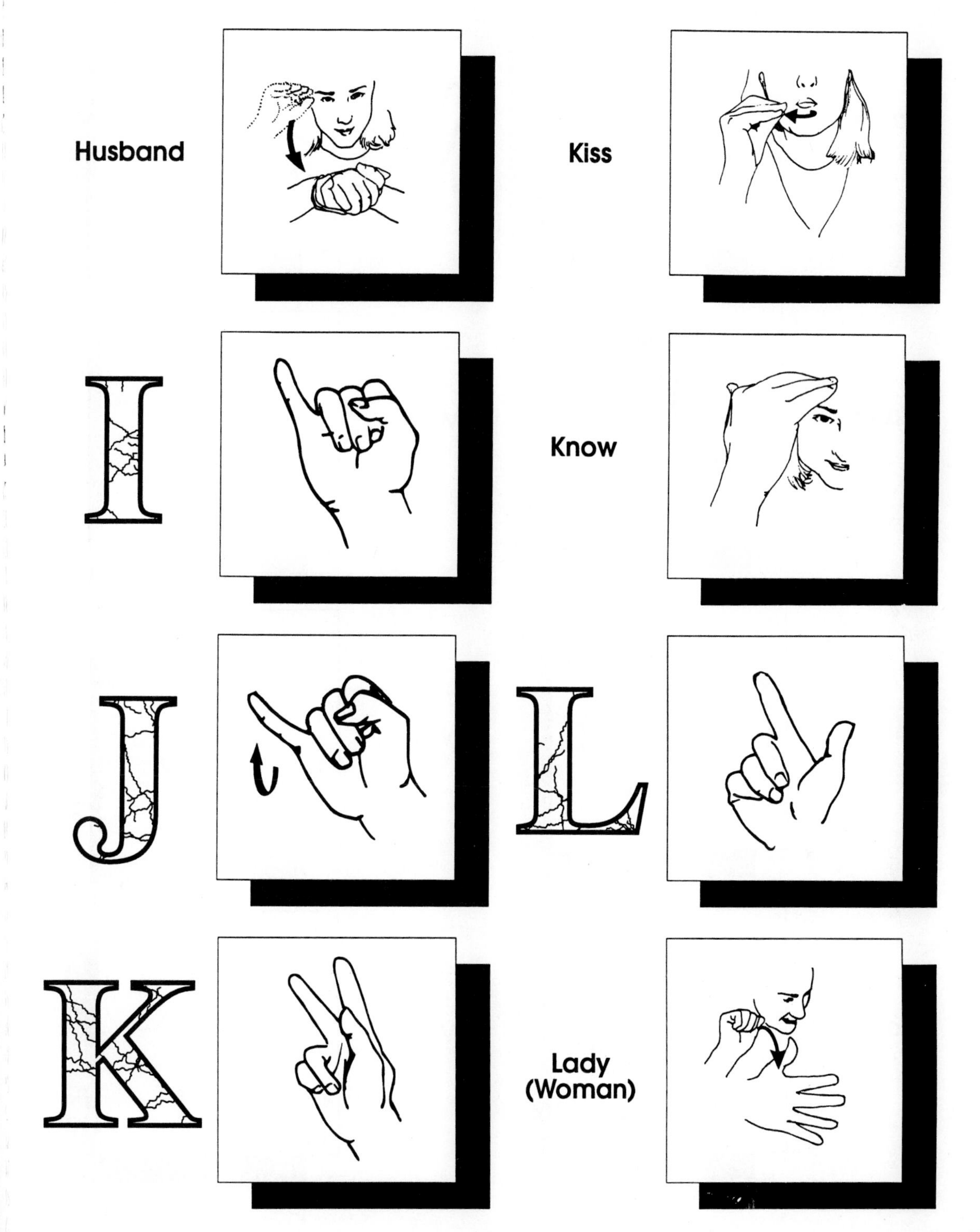
Husband
Kiss
I
Know
J
L
K
Lady
(Woman)

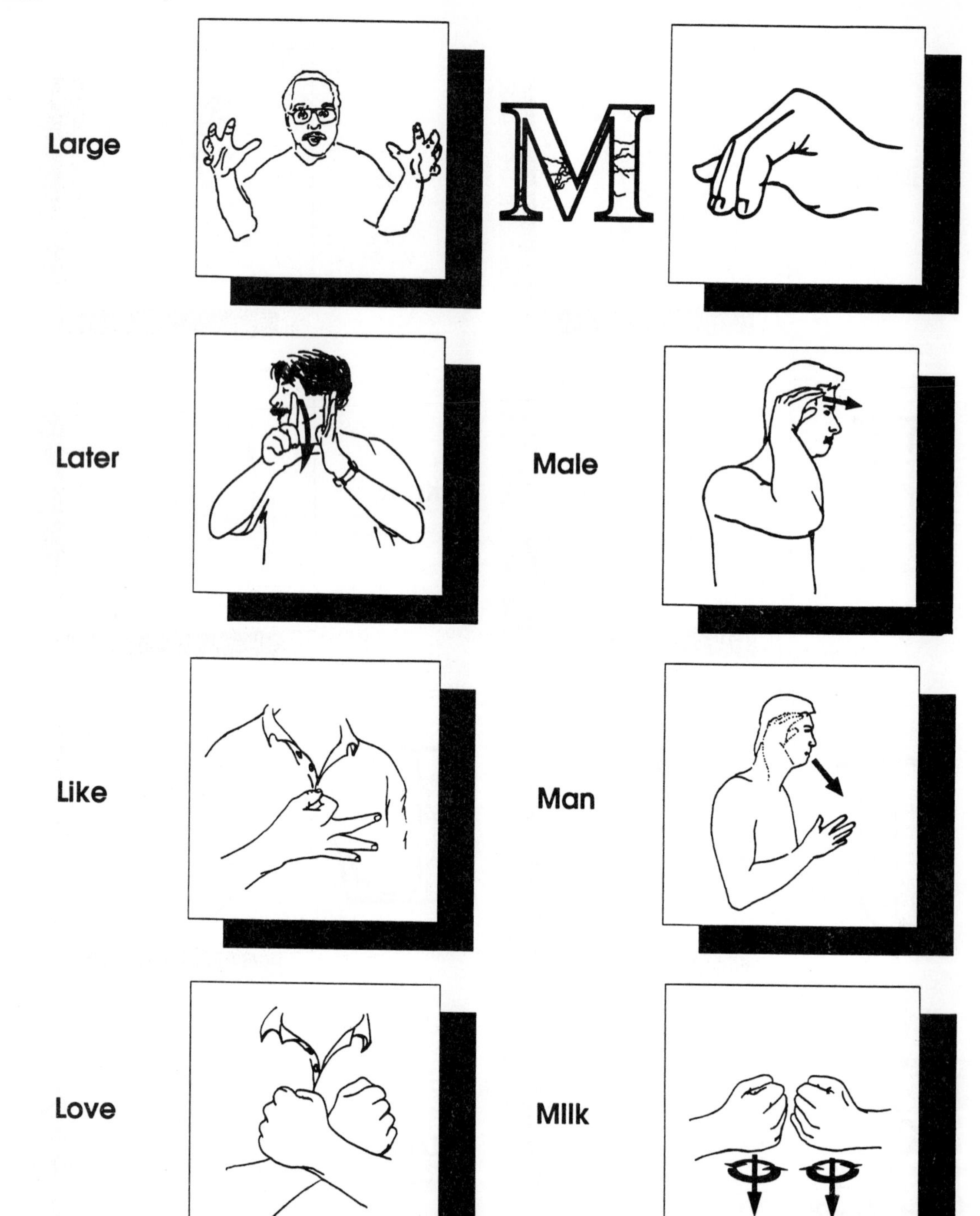
Large
M
Later
Male
Like
Man
Love
Milk

Mine

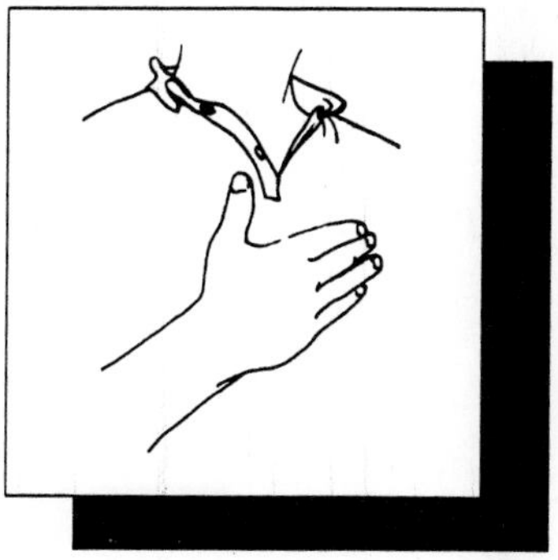

N

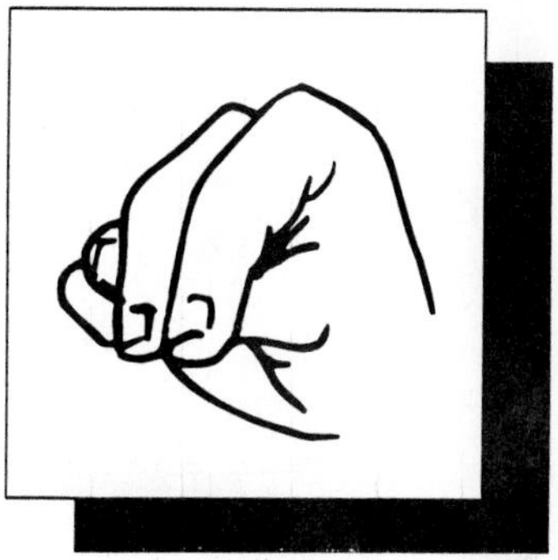

Morning

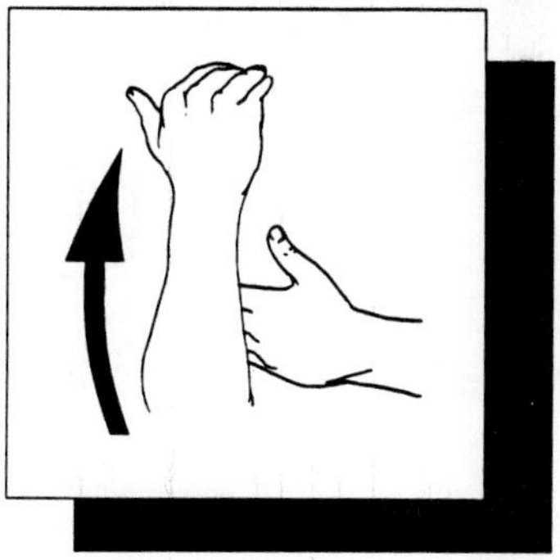

Name

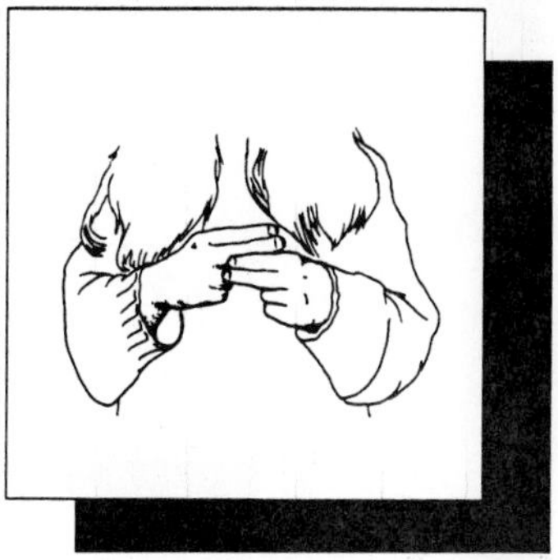

Mother

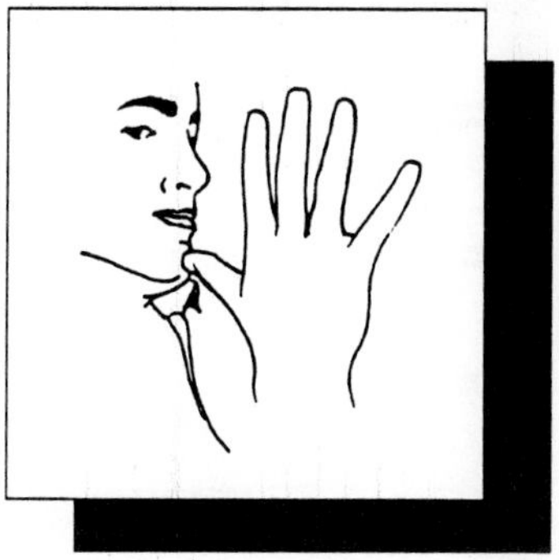

Narrow-minded

Music

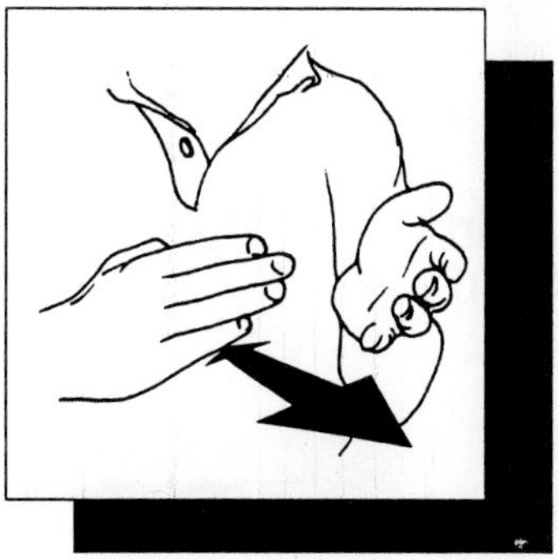

Necessary (Need)

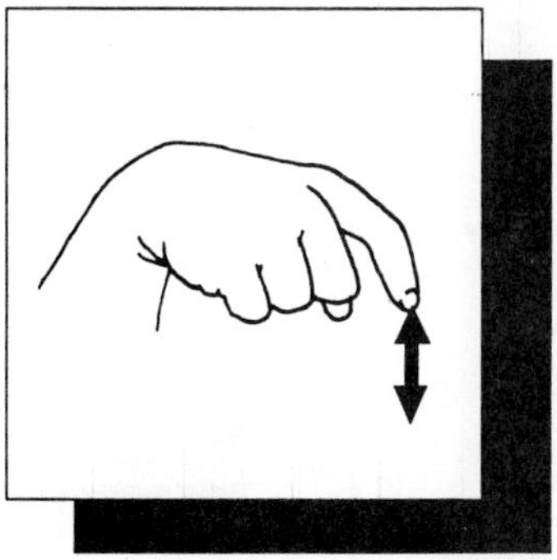

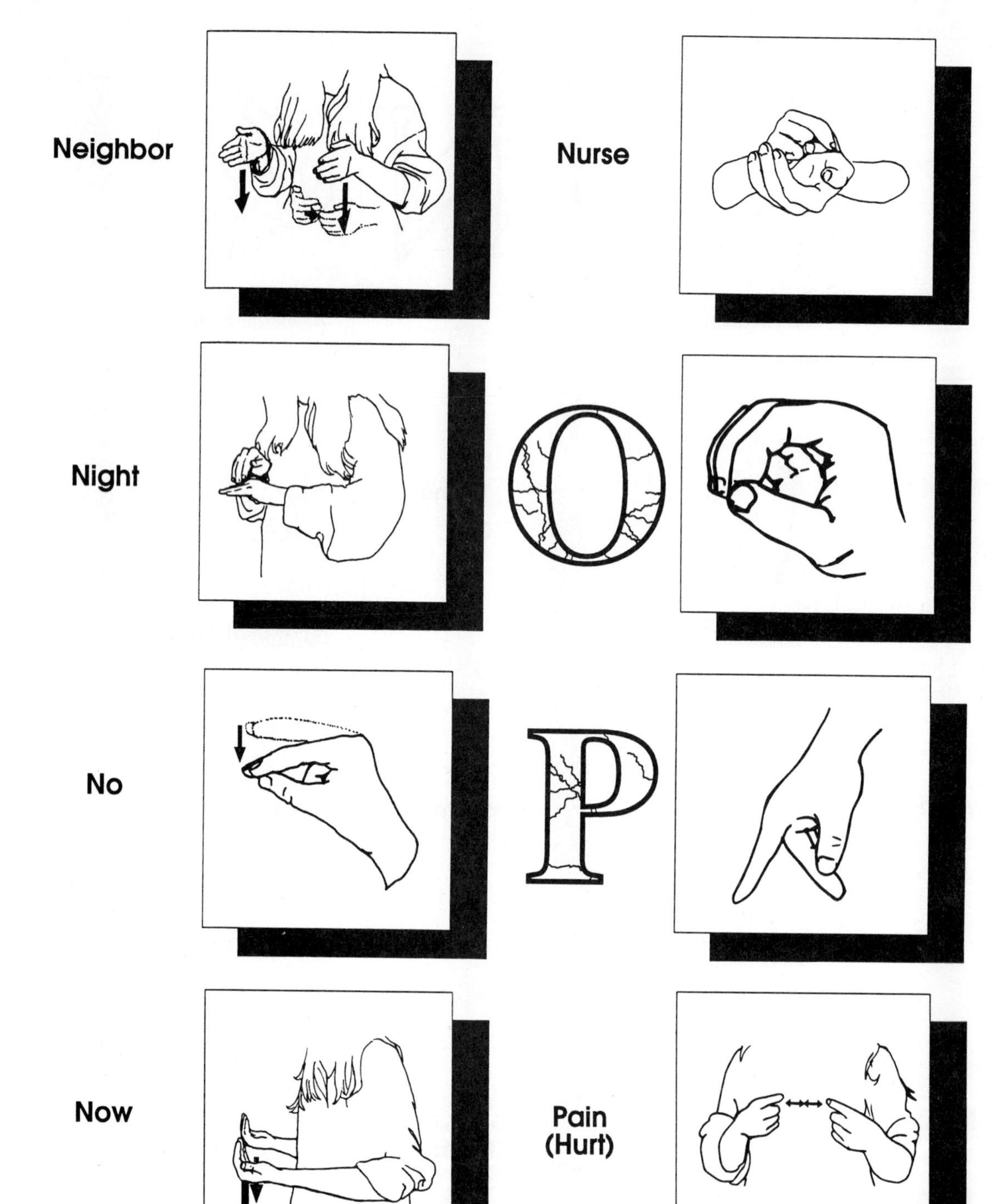
Neighbor
Nurse
Night
O
No
P
Now
Pain
(Hurt)

Parents

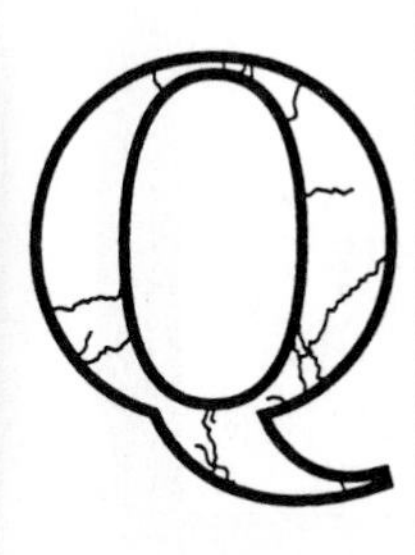

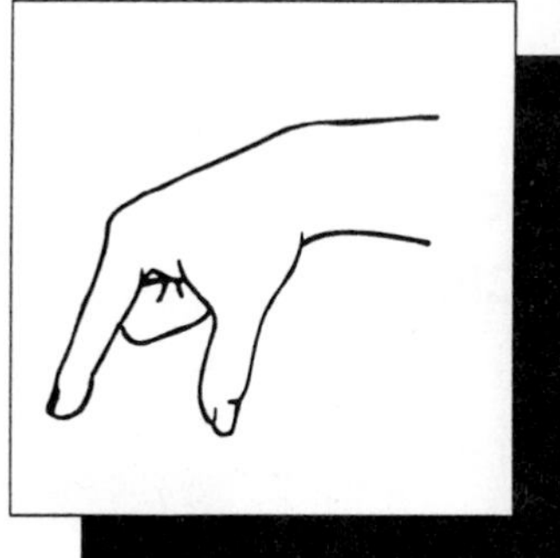

Person

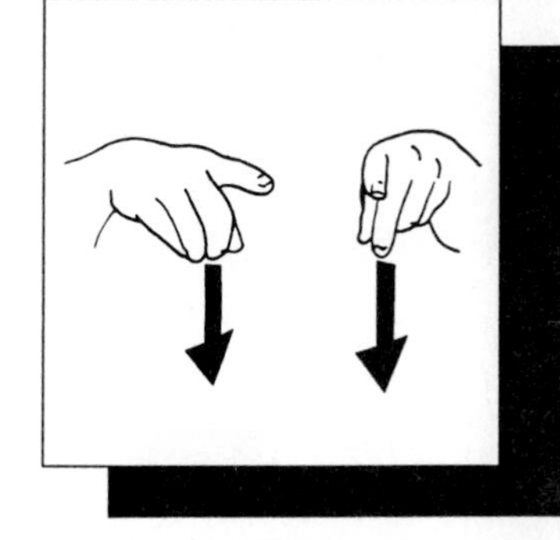

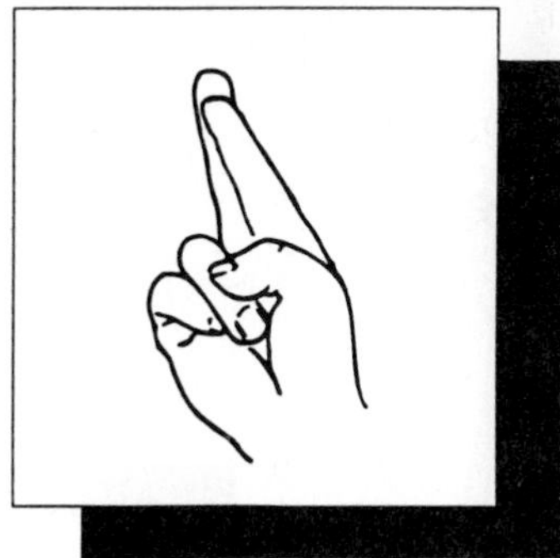

Please

Room
(Box)

Popcorn

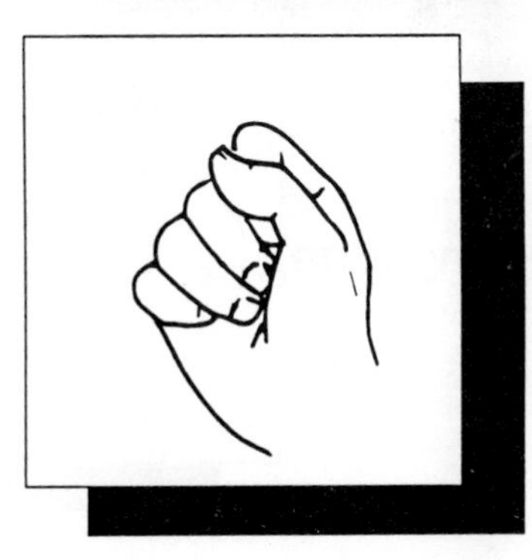

School
(Two Claps)
Similar
See
Sister
(Girl + Same)
Sick
Small
Sign
(Sign-Language)
Soda
(Soda Pop)

Sorry

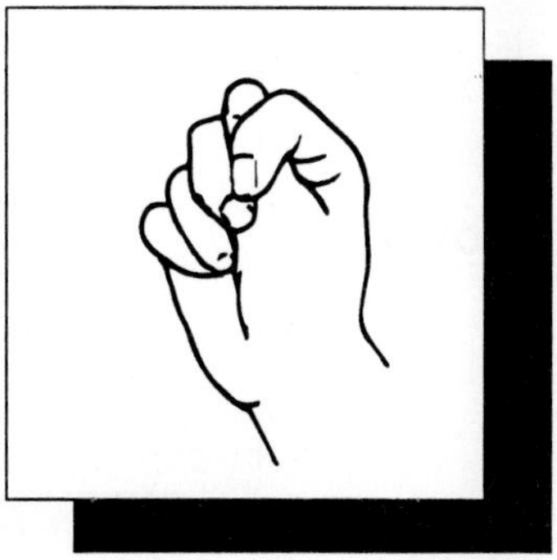

Take

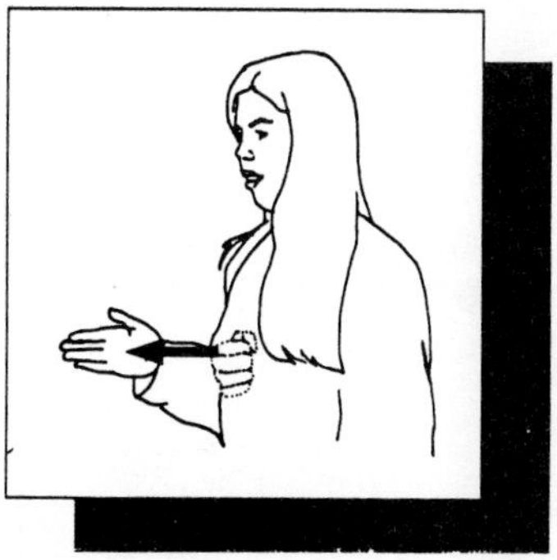

Spaghetti

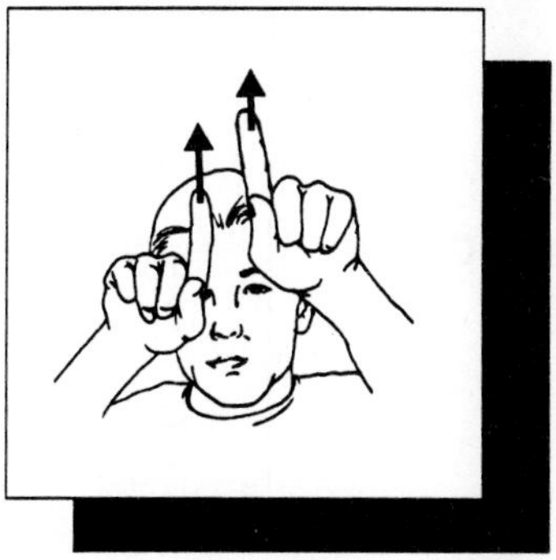

Talk

Star

Tea

T

Team

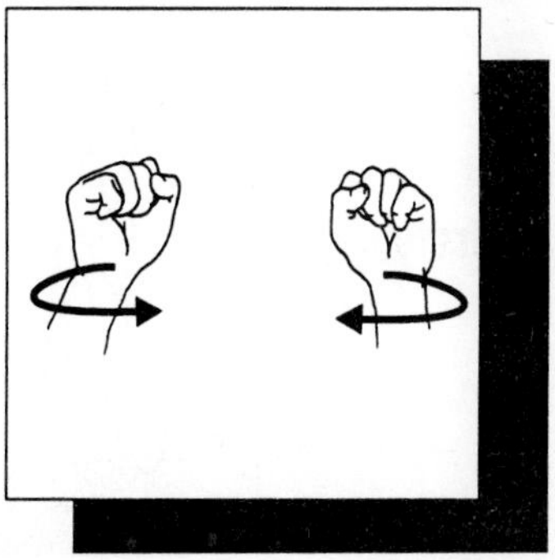

Tell
(Speak)
Those
Thank you
Today
Think
Tomorrow
Thirsty
Town

Tree

U

True

V

Try

W

Twice

Want

We

Week

What

When (1)

When (2)

Where

Whiskey

Who

Why

With

Wife

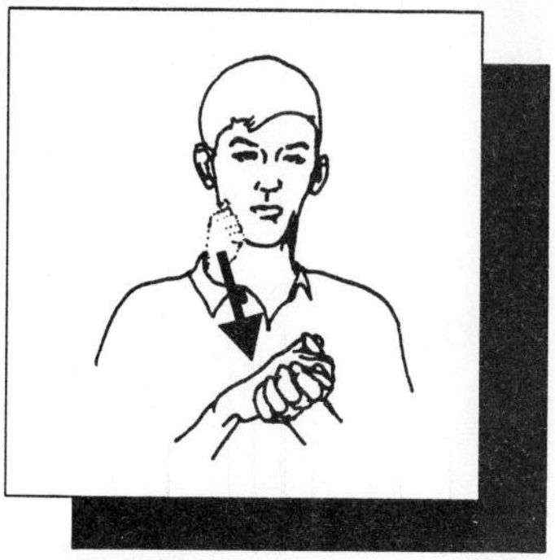

Word

Wine

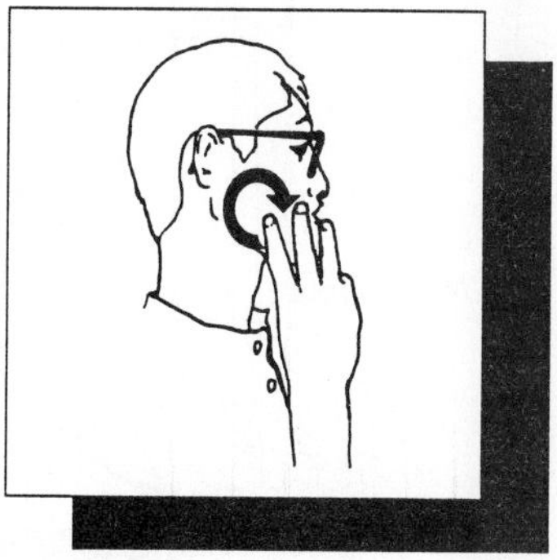

Work

Wisdom

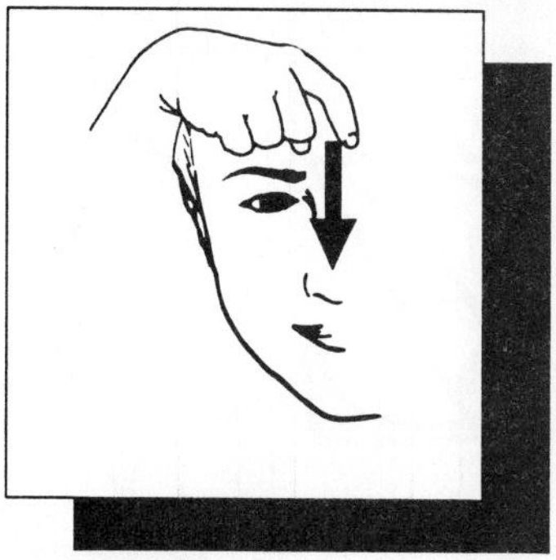

X

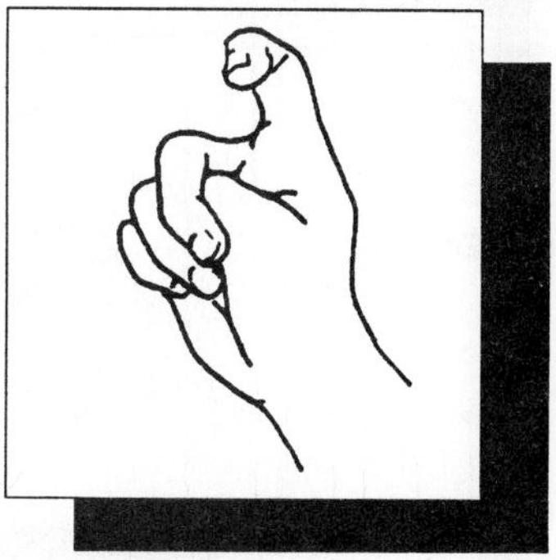

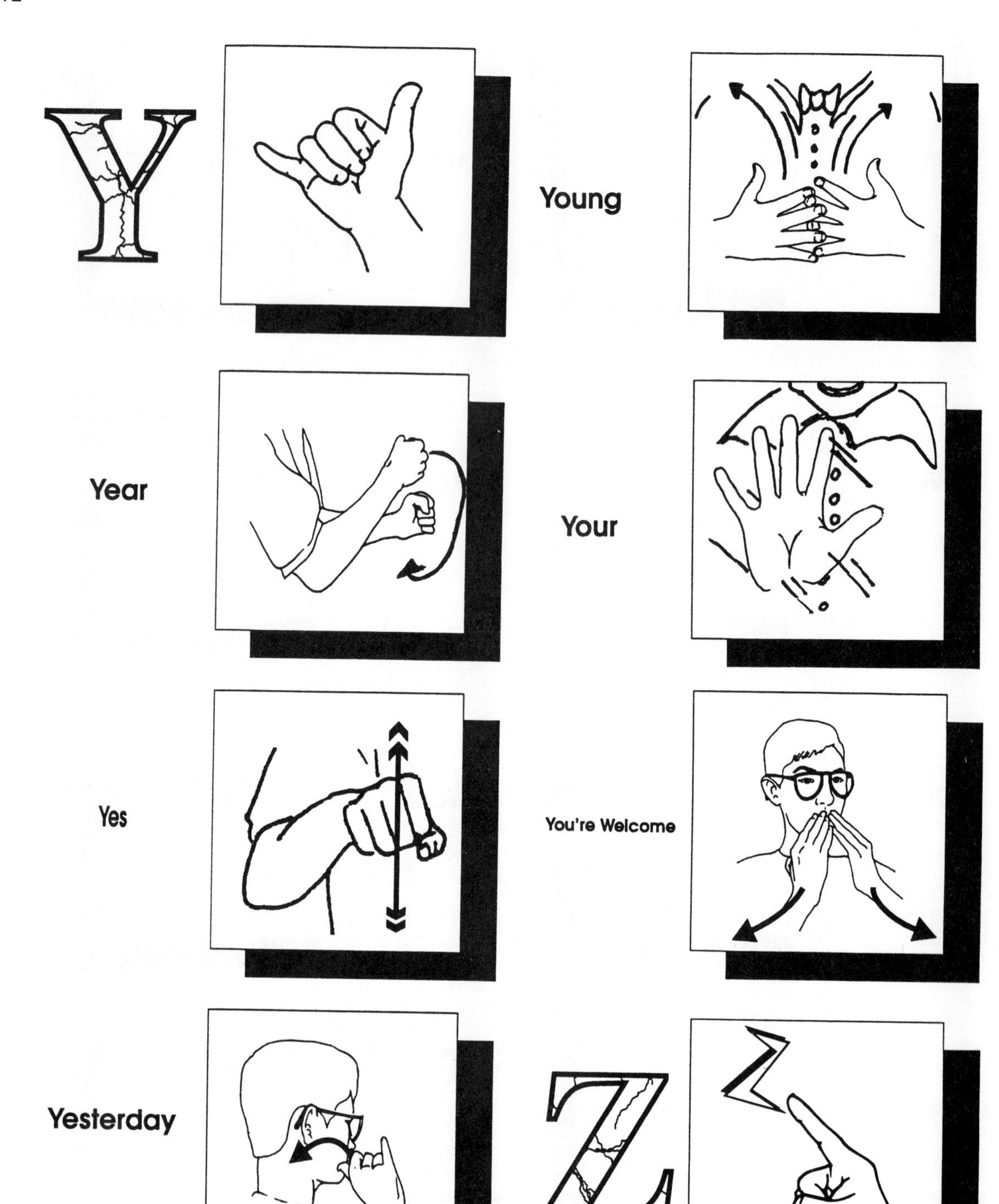
Y
Young
Year
Your
Yes
You're Welcome
Yesterday
Z

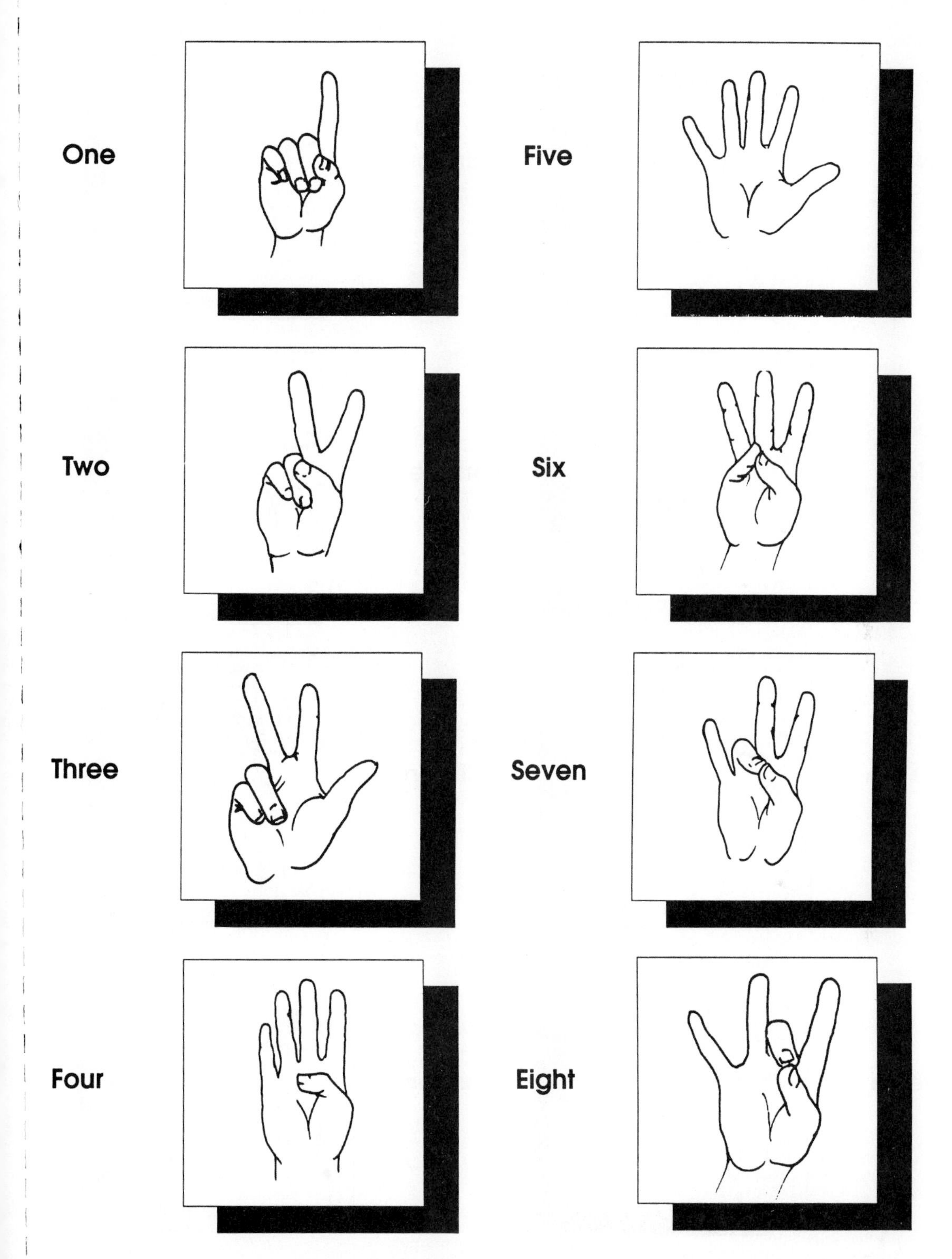
One
Two
Three
Four
Five
Six
Seven
Eight

Nine

Ten

Eleven

Twelve

Thirteen

Fourteen

Fifteen

Sixteen

Seventeen

Eighteen

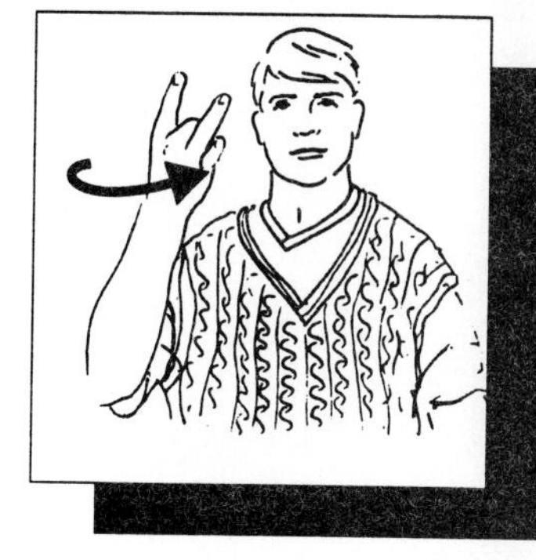

Nineteen

Twenty

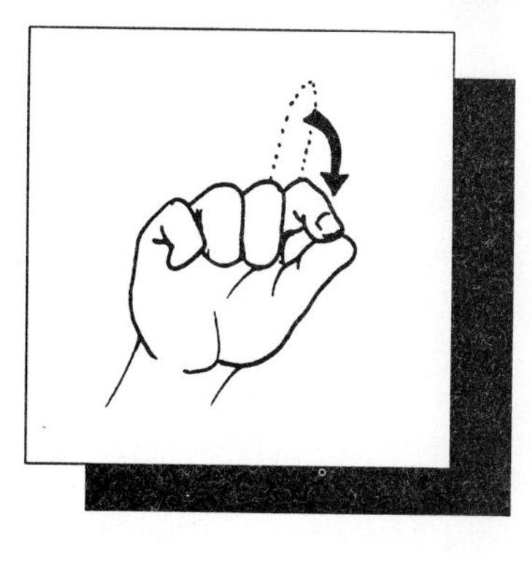

Twenty-One

One Hundred

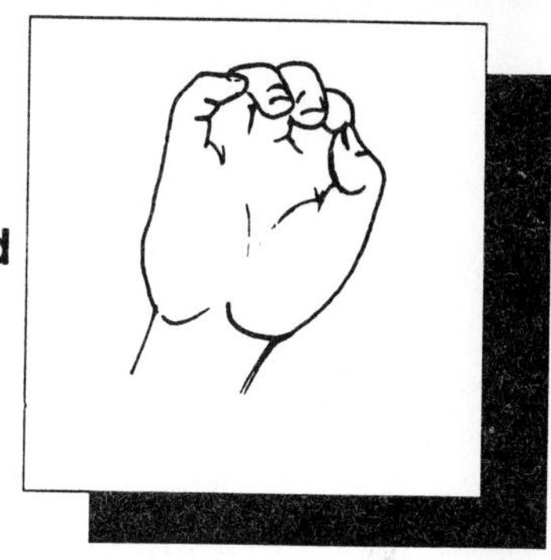

Thousand

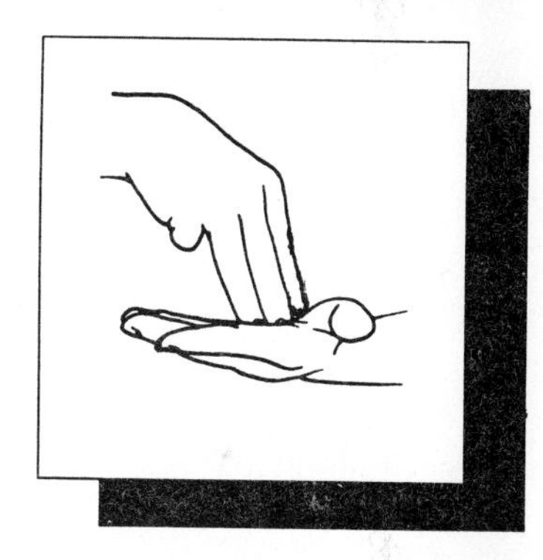

Million

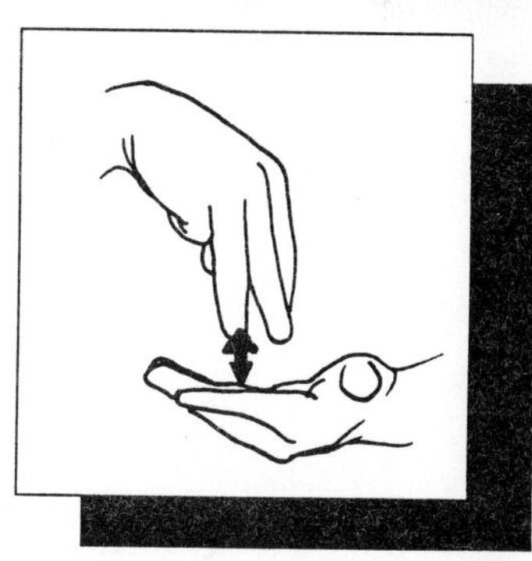

Notes

Human Diversity:
A Study Guide

by Craig Conley, B.S., M.A.

This study guide has been designed with three goals in mind:

- to review key points from each chapter and allow you to check your overall comprehension
- to prepare you for examinations by providing sample test questions
- to challenge you to consider some broader implications of your knowledge

Each chapter of this study guide corresponds to a chapter in the textbook. Four sections of exercises will help you to gain a fuller understanding of the concepts you just learned.

The first section, entitled **For Further Consideration**, builds upon the knowledge you gained from the chapter by encouraging you to determine how the material relates to your life.

Next, the **Terminology** section will give you the opportunity to review the new terms which were presented in the chapter. Try to define them from memory before looking back to the text for definitions.

Finally, the **True/False** and **Multiple Choice** sections examine some more specific concepts from the chapter. You can check your answers to these questions with the key located at the end of each set.

Chapter 1 Review

For Further Consideration

Look back at the quotation that opened chapter one. Intolerance, Helen Keller suggests, can be the greatest destructive force on earth. Yet where does tolerance begin? Which is more important—the individual or society? The one or the many? Unity or diversity? Ultimately, we must embrace both the individual and the multitude. We must integrate unity and diversity.

Diversity has to do with external qualities: human beings are diverse in that they are young and old; brown, tan, red, yellow, and pink; male and female; gay, straight, and bisexual; Christian, Jew, Buddhist, and Moslem; tall and short; large and small, rich and poor.

Unity, on the other hand, has to do with a more spiritual condition—the deeper nature of our thoughts and actions. Unity blossoms when we are peaceful, accepting, non-judgmental, and loving to ourselves and others.

As we recognize and respect the diverse qualities within ourselves, doors will open to creative relationships that would not be possible otherwise.

Check Yourself

Terms to Define:

1. exceptionality

2. labels

3. disability

4. diversity

True/False:

1. The expectations that a certain population has upon its members help to define exceptionality.

2. A disability is first and foremost a physical limitation.

3. Every person falls into some categor, of exceptionality.

4. Labeling may stigmatize someone and contribute to low self-esteem.

5. Labeling allows us to identify, distinguish, and describe people.

6. The expectations of influential people affect the success of exceptional people.

7. Even if you are in a bad mood, when you encounter someone who is different it is best to be pleasant, make eye contact, and smile.

8. When in the presence of a person who is different, you should substitute any possibly offensive vocabulary with less-offensive words.

9. Children should be discouraged from talking directly to people who are diverse about their differences.

10. We should treat all people the same, regardless of their differences.

Multiple Choice:

1. If you were the only person on earth not 8 feet tall, what would you prefer to be labeled?
   a. disabled
   b. handicapped
   c. exceptional
   d. vertically challenged

2. Whom should you not invite to go swimming with you?
   a. one who is mentally retarded
   b. one who is legally blind
   c. one with epilepsy
   d. one who can't swim

3. Why are labels important in our society?
   a. So we can identify people who are different.
   b. To compare the possible achievements of others with our own.
   c. So that professionals can communicate with each other about differences.
   d. Because people have an inherent need to classify and organize individuals within a societal group.

4. If someone has a negative reaction to your association with a person who is different, which of the following should you do?
   a. Say you were talking to a person, not a label.
   b. Say that you always try to be non-judgmental.
   c. Make it clear that you don't have anything in common with the exceptional person.
   d. both a and b.

5. When dealing with someone who is different, you should
   a. alter your expectations so you won't embarrass the person.
   b. forget labels and concentrate on the person.
   c. alter your behavior, being careful not to offend.
   d. all of the above.

6. Which of the following is not recommended?
   a. Say "It's good to see you again" to someone who is blind.
   b. Suggest going for a stroll through the park with someone in a wheelchair.
   c. Rush ahead to open the door for someone in a wheelchair.
   d. Expect normal achievement from someone with AIDS

<u>Answers</u>

1. T	1. any
2. F	2. d
3. T	3. c
4. T	4. d
5. T	5. b
6. T	6. c
7. F	
8. F	
9. F	
10. T	

Chapter 2 Review

For Further Consideration

While it may be important for the diverse individuals of the world to unite, the diverse cultures of these people do not have to be lost in the process. There is an important distinction between cultural unity and cultural *harmony.* Cultural unity means homogenizing all cultures into a single cultural unit. Cultural harmony, on the other hand, means all categories of cultural diversity remain unique but exist in accord with one another. Clearly, it is cultural harmony that we want to achieve. If we expect and respect cultural diversity, cultural harmony will blossom spontaneously.

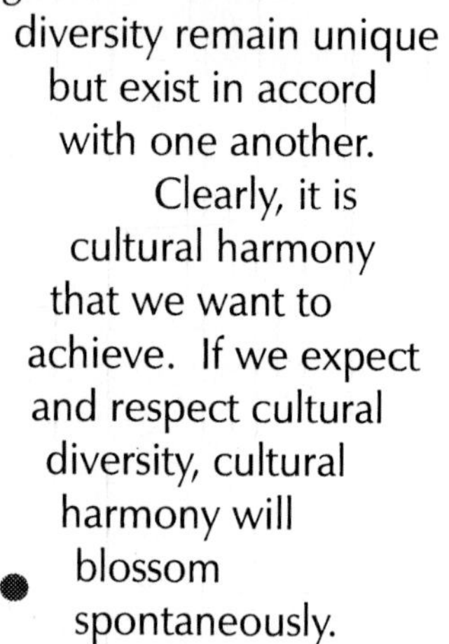

Society is like a big gumball machine. You never know what to expect. But you have to reach out your hand in order to grab the rewards.

Check Yourself

Terms to Define:

1. acculturation

2. ideal culture

3. real culture

4. explicit culture

5. implicit culture

6. ethnocentrism

7. melting pot

8. multicultural

9. subculture

10. cultural pluralism

True/False:

1. All cultures have a history, but not all have a heritage.

2. Group values are not technically part of a culture.

3. Rules of etiquette are social rather than cultural characteristics.

4. "Real" culture refers to what people say they believe.

5. "Ideal" culture refers to how people think they should behave.

6. Explicit culture can be described verbally.

7. Explicit culture includes religious beliefs.

8. "Acculturation" and "melting pot" refer to the same process.

9. Intercultural difficulties arise mainly from communication problems.

10. A microculture may also be a subculture.

Multiple Choice:

1. Which of the following does implicit culture *not* include?
   a. fears
   b. values
   c. tools
   d. hidden elements

2. Which of the following are aspects of "real" culture?
   a. behaviors
   b. stories
   c. proverbs
   d. jokes

3. The microculture seeks to
   a. distinguish itself from the larger political society.
   b. mediate the ideas, values, and institutions of the larger culture.
   c. occupy a subordinate position in society.
   d. none of the above.

4. Every culture is made up of:
   a. emotions
   b. interests
   c. slang expressions
   d. all of the above

5. Explicit culture does *not* include:
   a. fashions of dress
   b. speech
   c. tools
   d. assumptions

6. This refers to the belief that diverse groups coexist within society and maintain a culturally distinct identity:
   a. ethnocentrism
   b. intercultural
   c. pluralism
   d. melting pot

7. This refers to interactions which involve mutual or reciprocal cultural exchange:
   a. encapsulation
   b. intercultural
   c. pluralism
   d. melting pot

8. This refers to a closing off of one culture from others:

a. encapsulation
b. intercultural
c. microculture
d. melting pot

Answers

1. F	1. c
2. F	2. a
3. F	3. b
4. F	4. d
5. T	5. d
6. T	6. c
7. F	7. b
8. F	8. a
9. T	
10. T	

Chapter 3 Review

For Further Consideration

Look back at the quotation that opened chapter three. We may enact any number of laws, but it is ultimately up to each person, as Thomas Carlyle suggests, to become all that he or she can be. It is individual ideas, expressions, and sharing that generate and fuel our culture. Without the contributions of every member of society, our culture is impoverished. We must continue to work toward securing the full participation in society for each person, regardless of exceptionality.

Yet even as we envision a better tomorrow, we can begin changing the world today by improving our individual attitudes and dissolving our particular prejudices. It doesn't require legislation to treat our fellow human beings fairly. And once laws are passed, it takes individuals to see that they are carried out.

Check Yourself

Terms to Define:

1. equal protection

2. IDEA

3. ADA

4. Brown vs. the Board of Education

5. JTPA

6. 14th Amendment

7. Bill of Rights, Article I

True/False:

1. No well-defined body of law guarantees religious freedom.

2. State governments are prohibited from favoring certain groups of people over other groups.

3. No well-defined body of law guarantees educational opportunity.

4. Speech impairment is the only category of disability *not* covered by the Rehabilitation Act of 1973.

5. The 14th Amendment to the Constitution states that it is illegal to classify a citizen by his or her national origin.

6. Naturalized citizens are entitled to equal protection under the law.

7. The term "family" has a precise legal meaning.

Multiple Choice:

1. The Rehabilitation Act of 1973 protects the rights of:
   a. homosexuals
   b. religious institutions
   c. people with health problems or physical impairments
   d. older Americans

2. A "civil rights bill" for individuals with disabilities is the:
   a. Rehabilitation Act of 1973
   b. Education of All Handicapped Children Act of 1975
   c. Americans with Disabilities Act of 1990
   d. Equal Pay Act of 1963

3. The Equal Protection Clause of the 14th Amendment guarantees equal protection under the law for:
   a. the poor
   b. non-citizens
   c. prisoners
   d. all of the above

4. The Equal Protection Clause of the 14th Amendment does *not* guarantee the rights of:
   a. children
   b. minorities
   c. older Americans
   d. none of the above

5. No well-defined body of law guarantees the rights of:
   a. older Americans
   b. homosexuals
   c. low-income families
   d. both b and c

6. Which of the following terms does not have a precise legal meaning?
   a. family
   b. minority
   c. citizen
   d. religion

7. Which of the following types of laws insure educational opportunities?
   a. local
   b. state
   c. federal
   d. all of the above

Answers

1. F	1. c
2. T	2. c
3. F	3. d
4. F	4. d
5. T	5. b
6. T	6. a
7. F	7. d

Chapter 4 Review

For Further Consideration

It would be ideal to say that race doesn't matter, but we do not live in the best of all possible worlds. Realistically, race *does* matter to people and affects their feelings and actions in major ways, every day. There is only one solution to this situation: to redefine the importance we *ourselves* place on the concept of race. As we reformulate our own attitudes about race—one person at a time—society will start reflecting more enlightened perspectives.

Check Yourself

Terms to Define:

1. ethnicity

2. racism

3. scapegoating

4. prejudice

5. minority

6. discrimination

7. nationality

8. bias

9. bigotry

10. stereotyping

True/False:

1. "Americanization" and "melting pot" refer to the same phenomenon.

2. Ethnicity is primarily a product of perception.

3. Ethnicity and language are not related.

4. The genetic difference between two French women is greater than the difference between a French woman and a Spanish woman.

5. Human races are constantly changing and evolving.

6. There are only five genetically pure races in the world today.

7. There are as many as 2,000 completely distinct races.

8. Humans have always divided themselves into racial categories.

9. The concepts of race and culture are inseparable.

10. The gene that governs racial differences has been isolated by scientists.

Multiple Choice:

1. We use racial categories to describe an individual's:
    a. personality
    b. intelligence
    c. character
    d. body build

2. A person's ethnicity has to do with his or her:
    a. race
    b. country of origin
    c. culture
    d. none of the above

3. A personal preference that prevents one from making a fair judgment is called a:
    a. prejudicial attitude
    b. bias
    c. discriminatory practice
    d. none of the above

4. The policy of blaming a person or group when the fault lies elsewhere is:
    a. bigotry
    b. discrimination
    c. scapegoating
    d. bias

5. According to the Census Bureau, one's race is determined according to:
    a. personal preference
    b. country of origin
    c. parents' genetic stock
    d. parents' country of origin

6. The concept of a "minority" is derived from:
    a. anthropology
    b. sociology
    c. genetics
    d. none of the above

7. The concept of race is derived from:
    a. anthropology
    b. sociology
    c. genetics
    d. none of the above

Answers

1. T	1. d
2. T	2. c
3. F	3. b
4. T	4. c
5. T	5. a
6. F	6. b
7. F	7. a
8. F	
9. F	
10. F	

Chapter 5 Review

For Further Consideration

We might consider that we have conflict today *because* of the opportunity not to go on doing the same old things and thinking in the same old ways. Conflicts create the pressure to change. It is also important to remember that good communication may not create agreement. Decide whether you want civility or true understanding of your viewpoint. How much damage to yourself are you willing to risk? You are entitled to respect, but you are not entitled to acceptance of your point of view, necessarily.

Here are some pointers for conducting a civil discussion about gender and sexual orientation: Take responsibility for maintaining civil dialogue. Focus on commonalities. Don't tolerate degrading, demeaning, or hurtful attitudes. Don't squelch differences. Don't run away from emotions.

Check Yourself

Terms to Define:

1. chauvinism
2. sexism
3. homosexuality
4. gender gap
5. homophobia
6. bisexual
7. nationality
8. bias
9. bigotry
10. stereotyping

True/False:

1. All people suffer from gender stereotypes.

2. Scientists agree that people are born either heterosexual or homosexual.

3. A person's sexual orientation may change over the course of his or her life.

4. Gay men and women may be married to opposite sex partners and have children.

5. Women who drive trucks or work for the military are likely to be lesbians.

6. Any literature having to do with sexuality is pornography.

7. The way a person dresses, talks, and behaves can reveal his or her sexual orientation.

8. Men as well as women are victims of gender inequality.

9. Some sexual orientations are more "normal" than others.

10. Sexual activity and sexual orientation mean the same thing.

<u>Multiple Choice:</u>

1. Which of the following occupations could be typical of a gay man?
   a. truck driver
   b. interior decorator
   c. athlete
   d. all of the above

2. Which of the following issues is a concern of the men's movement?
   a. child custody rights
   b. suicide prevention
   c. homelessness
   d. all of the above

3. The belief that one gender is superior to another is called:
   a. homophobia
   b. sexism
   c. chauvinism
   d. discrimination

4. Which of the following terms is *not* related to sexual orientation:
   a. homosexual
   b. transsexual
   c. transvestite
   d. none of the above

5. Which of the following statements is false?
   a. The majority of the population is strictly heterosexual.
   b. Reproductive rights are only one concern of the women's movement.
   c. It is best to call people what they prefer to be called.
   d. Labeling people promotes divisiveness.

6. Gender identity is formed by:
   a. parents
   b. peers
   c. society
   d. all of the above

7. Pedophiles are males and females who are
   a. homosexual
   b. heterosexual
   c. bisexual
   d. none of the above

8. Transvestites wear clothing of the other gender and are predominantly

a. homosexual
b. heterosexual
c. bisexual
d. transsexual

Answers

1. T	1. d
2. F	2. d
3. T	3. c
4. T	4. b
5. F	5. a
6. F	6. d
7. F	7. d
8. T	8. b
9. F	
10. F	

Chapter 6 Review

For Further Consideration

Mutual respect is the key to getting along with people who do not share your faith. Allow each other to express religious beliefs honestly, without fear of rejection. Accept what the other person says. You may not agree with him or her, but you can demonstrate that you accept that person's feelings. You show acceptance through the tone of voice and the words you use. Be a reflective listener. Make eye contact and concentrate on what the other says. Don't feel compelled to respond. Silently listen and make the person feel understood. Be a mirror for that person to see himself or herself more clearly.

Check Yourself

<u>Terms to Define:</u>

1. persecution

2. religion

3. karma

4. nirvana

5. Allah

6. Yahweh

7. meditation

8. reincarnation

9. Agnosticism

10. Shamen

11. Atheism

<u>True/False:</u>

1. All faiths share the same underlying religious impulse.

2. In order to understand another faith, you must believe its doctrines.

3. The *Talmud* collects the wisdom of Hinduism.

4. Judaism, Christianity, and Islam all worship the same God.

5. Judaism, Christianity, and Islam all acknowledge the same prophets.

6. Shinto is the dominant religion in Taiwan.

7. There is no such thing as a Confucian church.

8. An atheist is a skeptic.

9. New Religions are concerned primarily with social order.

10. Witchcraft is an ancient religion.

Multiple Choice:

1. A prophet is a:
   a. teacher
   b. visionary
   c. messenger
   d. all of the above

2. Tribal societies in Australia are typically:
   a. atheist
   b. animist
   c. adventist
   d. none of the above

3. Reincarnation is a belief of:
   a. Sikhism
   b. Buddhism
   c. Hinduism
   d. all of the above

4. Sikhism grew out of:
   a. Confucianism
   b. Jainism
   c. Islam and Hinduism
   d. Confucianism and Animism

5. Taoism teaches people to be:
   a. courageous
   b. peaceful
   c. loyal
   d. passionate

6. All religions
   a. answer human questions.
   b. promote national loyalty.
   c. claim to be favored by God.
   d. stress self-discipline.

7. Studying other religions
   a. strengthens your own faith.
   b. helps you to see your faith in a universal context.
   c. improves human relations.
   d. all of the above

8. Which religion did *not* originate in the Middle East?
   a. Baha'i
   b. Sufism
   c. Jainism
   d. Zoroastrianism

Answers

1. T	1. d
2. F	2. b
3. F	3. d
4. T	4. c
5. F	5. b
6. F	6. a
7. T	7. d
8. F	8. c
9. T	
10. T	

Chapter 7 Review

For Further Consideration

What would you do if you had all the money in the world? When a young child was asked that question, the answer was, "Spend it." That seemed at first to be a childish and simplistic answer, but upon reflection it seems to be the only one. Money is just paper unless it circulates. It is natural to want all the money in the world. We are all entitled to the earth's bounty, and naturally we want to enjoy the freedom and the material comforts that money can buy.

Few of us would want to keep all the money in the world locked away in a safe. It would do nothing for us. It could not bring us pleasure nor could we use it as a tool for making more money. The question, then, becomes: Spend it on what? We can learn a lot about our real priorities by coming up with an honest answer. Will you travel? Give to a particular charity? Support a certain cause? Fund a scholarship in a field of study? Shower gifts on friends and family? Andrew Carnegie established libraries. Rockefeller endowed the arts. Thinking about how you would spend all the money in the world helps you to clarify your idea of money itself. Do you see it as a seed to be sown, a voucher for future security to be hoarded, or as one more element of the universal flow, to be channeled and used and passed along as you follow your life path?

Oscar Wilde was being facetious when he said the quotation that opens this chapter, but there is an element of truth to it. Try substituting the word "affluence" for money. Being affluent doesn't just mean being rich in dollars. It means being rich in generosity as well. As Dr. Deepak Chopra explains in his book *Creating Affluence,* a truly wealthy person never focuses his or her attention on money alone. Whatever the size of your bank account, if money is all you can think about then you are ultimately impoverished because you never feel satisfied. Allow your kindness, generosity, and good will to flow freely and you will increase affluence in your own life and in the people around you.

Check Yourself

Terms to Define:

1. socioeconomic

2. underclass

3. wealth

4. working class

5. social status

6. prestige

True/False:

1. Most residents of poverty areas are not poor.

2. More than half the people living in poverty areas are African Americans.

3. Labeling and stereotyping in regard to socioeconomic status are as harmful as any other labeling and stereotyping.

4. People often confuse zip codes with socioeconomic status.

5. More Americans than not live in poverty areas.

6. The United States has a lower incidence of poverty than Norway, West Germany, and Japan.

7. The underclass is so-named because it lies on the outskirts of the class system.

Multiple Choice:

1. Socioeconomic status refers to a person's
   a. income
   b. education level
   c. occupation
   d. all of the above

2. The poverty rate for children under 18 is approximately
   a. 12%
   b. 22%
   c. 32%
   d. 42%

3. Prestige may be acquired through
   a. occupation
   b. appearance
   c. dress
   d. all of the above

4. Which group makes up the majority of people living in poverty?
   a. families with a female head of household
   b. married couples
   c. recent immigrants
   d. children born in another country

5. The largest segment of society falls into the
   a. lower class
   b. upper class
   c. lower middle class
   d. underclass

6. Social class is determined by one's
   a. income
   b. influence
   c. family name or bloodline
   d. all of the above

7. The majority of families living in poverty are
   a. Non-Hispanic Caucasian
   b. African American
   c. Asian and Pacific Islander
   d. Hispanic

8. Approximately what percentage of African Americans live in poverty?
   a. 13%
   b. 23%
   c. 33%
   d. 43%

<u>Answers</u>

1. T	1. d
2. F	2. b
3. T	3. d
4. T	4. a
5. T	5. c
6. F	6. d
7. T	7. a
	8. c

Chapter 8 Review

For Further Consideration

You are more than your physical body. Many of your best qualities are invisible, such as your personality, imagination, and ability to love. But your physical body is the first thing others notice about you. When you meet a stranger, you can't help but notice if he or she is short or tall, thick or thin. You may notice the person's skin color and haircut. You may notice the color of the eyes. These physical qualities tell you about the person's looks, but nothing about the personality underneath. You would be more likely to notice visible aspects of people with physical differences than you would be when observing individuals with learning, sensory, sexual orientation, health, religious, or intellectual differences.

Most people would like to change at least one thing about their appearance. Some people go to a surgeon to get a smaller nose. Other people sunbathe to get darker skin. Everyone tries out different hair styles. Many people join health clubs to develop their muscles or get in shape. Many others go on special diets to lose or gain weight. A person's physical qualities can be changed, and everyone's physical body changes as they grow older.

People with physical differences can, and do, make significant contributions to society. Most people with physical differences would like greater understanding from others and are willing to educate others about their physical differences. In order for this to occur, however, you have to display the desire to learn. Next, you need to interact directly with people with physical differences. You can learn a great deal from their example.

Check Yourself

<u>Terms to Define:</u>

1. morphology

2. gigantism

3. dwarfism

4. orthopedic impairment

5. physical disability

6. Cerebral Palsy

7. multiple sclerosis

8. Muscular Dystrophy

9. paraplegia

True/False:

1. Scientists have isolated a "fat gene" which can predispose a person to be overweight.

2. Anorexia refers to a person genetically predisposed to thinness.

3. The pituitary gland produces hormones which stimulate body growth.

4. Oversecretion of the pituitary gland produces dwarfism.

5. One in twenty babies is born with some type of disability.

6. Physical disabilities are problems caused by injuries or conditions of the central nervous system which interfere with mobility, coordination, communication, or behavior.

7. Quadriplegia refers to paralysis of the lower part of the body.

8. Spasticity is a wasting away of muscles which are unused.

Multiple Choice:

1. Persons who are obese are discriminated against in
   a. education
   b. employment
   c. public accommodations
   d. all of the above

2. This refers to sudden, uncontrollable muscle contractions:
   a. prosthesis
   b. atrophy
   c. paraplegia
   d. spasticity

3. A physical disability may involve problems with
   a. mobility
   b. coordination
   c. communication
   d. all of the above

4. A limiting condition a person has from birth is called
   a. spasticity
   b. atrophy
   c. congenital
   d. orthopedic

5. The major cause of paraplegia in young children is
   a. Spina Bifida
   b. Cerebral Palsy
   c. Multiple sclerosis
   d. Muscular Dystrophy

6. Paralysis of one side of the body is called
- a. paraplegia
- b. hemiplegia
- c. monoplegia
- d. quadriplegia

7. In addition to muscle weakness, spasticity, and balance difficulties, multiple sclerosis can also cause:
- a. dwarfism
- b. numbness of the appendages
- c. immune system deficiencies
- d. pulmonary difficulties

8. This has to do with the bones, muscles, and joints used in movement:
- a. orthopedic
- b. atrophy
- c. congenital
- d. spasticity

9. Paralysis of one extremity is called
- a. paraplegia
- b. hemiplegia
- c. monoplegia
- d. quadriplegia

Answers

1. T	1. d
2. F	2. d
3. T	3. d
4. F	4. c
5. T	5. a
6. T	6. b
7. F	7. b
8. F	8. a
	9. c

Chapter 9 Review

For Further Consideration

William Shakespeare said that "Learning is but an adjunct to ourself." He meant that what we learn and how we learn does not affect or alter our true selves. Yet the area of learning disabilities has generated more controversy, confusion, and polarization among contemporary professions than any other area of exceptionality. Typically, children with learning disabilities have normal intelligence, but they experience academic difficulties, and perhaps social problems as well.

Although discrepancies in prevalence estimates exist in all areas of exceptionality, the area of learning disabilities seems more variable than most, and it is one of the largest categories among exceptionalities. It is important to understand the varieties of learning difficulties, but at the same time we should remember Shakespeare's words and not confuse the disability with the person.

Check Yourself

Terms to Define:

1. dyslexia
2. hyperactivity
3. dyscalculia
4. dysgraphia
5. attention deficit disorder
6. aphasia
7. specific learning disability
8. perception disorders
9. aphasia

True/False:

1. Because a learning disability is invisible, the problem is not as severe as a physical disability.

2. No two people learn in exactly the same way.

3. Most people use technology to augment their learning.

4. Twenty percent of college students have been identified as having learning disabilities.

5. By definition, an individual with a learning disability has a lower than average IQ.

6. Learning disabilities frequently inhibit scholastic achievement, but they rarely inhibit social development.

7. Poor motivation has been identified as a key problem with many people diagnosed with learning disabilities.

8. Of all categories of exceptionality, learning disabilities require the greatest number of personnel.

Multiple Choice:

1. Learning experts estimate that what percent of our population has a learning disability?
   a. 1-15%
   b. 15-50%
   c. 50-75%
   d. 75-85%

2. Which of the following terms is used by psychologists to describe a learning disability?
   a. perceptual disorder
   b. aphasia
   c. brain injury
   d. specific learning disability

3. Which of the following terms is used by educators to describe a learning disability?
   a. perceptual disorder
   b. dyslexia
   c. brain injury
   d. specific learning disability

4. Which of the following terms is used by speech and language specialists to describe a learning disability?
   a. hyperkinetic disability
   b. dyslexia
   c. brain injury
   d. specific learning disability

5. Which of the following terms is used by doctors to describe a learning disability?
   a. hyperkinetic disability
   b. brain injury
   c. aphasia
   d. specific learning disability

6. Which of the following may be considered a "specific learning disability"?
   a. dyscalculia
   b. dyslexia
   c. both a and b
   d. none of the above

7. A person who counts on fingers and enjoys non-stationary activities may be a
   a. tactile/kinesthetic learner
   b. auditory learner
   c. visual learner
   d. slow learner

8. A person who moves lips while reading and seldom writes things down may be a

a. tactile/kinesthetic learner
b. auditory learner
c. visual learner
d. slow learner

9. A person who daydreams and has difficulty during lectures may be a

a. tactile/kinesthetic learner
b. auditory learner
c. visual learner
d. slow learner

Answers

1. F	1. b
2. T	2. a
3. T	3. d
4. F	4. b
5. F	5. b
6. F	6. c
7. T	7. a
8. T	8. b
	9. c

Chapter 10 Review

For Further Consideration

As with any category of human diversity, there is no one consistent type or profile of people with intellectual differences. The most gifted individuals to people with severe mental retardation span socioeconomic barriers, genders, races, and disabilities. People with either the highest or lowest intellects of any age benefit from special attention. Such attention may be provided in the home, the school, the workplace, and in leisure settings. Societal perceptions, stereotypes, expectations, and interactions affect the degree to which a person who is retarded or gifted may perceive his or her own abilities. As with any exceptionality, the important issue is that everyone is a person, deserving of the same respect and opportunities we would expect for ourselves.

Check Yourself

Terms to Define:

1. developmental delay

2. giftedness

3. Down syndrome

4. intelligence

5. phenylketonuria

True/False:

1. Gifted people by definition have advanced intellectual abilities.

2. Persistence is a prominent characteristic of gifted people.

3. People with mental retardation generally need to be institutionalized.

4. Developmental delays are most often identified before the child enters school.

5. Down syndrome is caused by a chromosomal abnormality.

6. PKU is a hereditary condition.

7. Results of intelligence tests are not affected by cultural differences.

8. Life skills and mechanical aptitude are considered when defining a person's intelligence.

Multiple Choice:

1. Which of the following is not a known cause of developmental delays?
   a. PKU
   b. CVB
   c. toxic agents
   d. Down syndrome

2. Valid assessment of developmental delays considers:
   a. cultural diversity
   b. linguistic diversity
   c. communication differences
   d. all of the above

3. Based upon census data, mental retardation directly affects how many people?
   a. 1 in 10
   b. 1 in 20
   c. 1 in 30
   d. 1 in 50

4. Developmental delays caused during the birth of the child are called
   a. postnatal
   b. prenatal
   c. perinatal
   d. psychosocial

5. It is safe to say that most gifted and talented people
   a. make good judgments.
   b. do not have to work hard in order to succeed.
   c. excel in more than one area.
   d. none of the above

6. The two most commonly used intelligence tests are:
   a. Stanford Binet Intelligence Scale and Weschler Intelligence Scale for Children
   b. Kaufman Assessment Battery and Guilford Model
   c. System of Multicultural Pluralistic Assessment and Guilford Model
   d. Trial and Error Intelligence Scan and SDGV Method

7. Developmental delays caused by malnutrition, lack of stimulation, poor medical care, and other environmental causes are called
   a. postnatal
   b. prenatal
   c. perinatal
   d. psychosocial

Answers

1. F	1. b
2. T	2. d
3. F	3. a
4. F	4. c
5. T	5. d
6. T	6. a
7. F	7. d
8. T	

Chapter 11 Review

For Further Consideration

Imagine that you acquired a serious disease and have only six months left to live. What will you do? How will you spend your time? Will your life have more authenticity for your acknowledging that it is limited? Perhaps you have devoted most of your time to your career. Will you now spend your time at home with your family? Or is your work so compelling and so important that you will continue it to the end? These questions help you to clarify your real priorities

It is unlikely that most of us would give up our hobbies if our time were limited to six months. If we love painting, we would paint. If we enjoy sailing, we would sail. All those things which enrich our lives and give meaning to them would probably take priority. So why is it that we require a catalyst such as a death sentence to force us to commit to doing what we find most rewarding and enjoyable? We don't have to wait. We can choose to take the risk and embrace life.

Now that you have dealt with the news of your limited life span, consider another scenario. A wonder substance has been discovered. It is being added to water supplies all over the world. After today, the average lifespan is 600 years. That calendar in your mind suddenly is more than seven times as long as you had previously constructed it. 600 sounds like all the time in the world, doesn't it?

Continue the exercise your started earlier. How does knowing that you will live 600 years affect your life choices? Where will you invest your energy, knowing that you have so much more than you'd imagined? Which endeavors seem relevant over this long term? Does your stake in the world seem different? Is your interest in the environment increased? What books do you now feel you have time to read? What good deeds do you have time to perform for your neighbor now that the pressures of time have been lifted? Will you do nothing, feeling that there's no hurry? Or will you soon adjust your mind so that 600 years seems short?

Check Yourself

<u>Terms to Define:</u>

1. asthma

2. diabetes

3. epilepsy

4. hemophilia

5. leukemia

6. sickle-cell anemia

7. tuberculosis

8. AIDS

True/False:

1. AIDS can be prevented.

2. HIV is spread through sharing bodily fluids.

3. Practicing safe sex prevents the spread of AIDS.

4. Practicing safe sex merely reduces the spread of AIDS.

5. By definition, diabetes involves an insulin dependency.

6. All cancers caused by smoking could be prevented entirely.

7. Asthma affects almost 10 percent of children in the U.S.

8. Episodes of asthma can be triggered by emotional factors.

9. A symptom of diabetes is periodic seizures.

10. Tuberculosis is a communicable disease.

Multiple Choice:

1. What type of disorder is asthma?
   a. central nervous system
   b. metabolic
   c. respiratory
   d. circulatory

2. AIDS is viral in nature and acts by attacking and weakening the body's:
   a. immune system
   b. reproductive system
   c. neurological system
   d. glandular system

3. Health impairments that are treatable but incurable are called:
   a. acute
   b. extended
   c. chronic
   d. orthopedic

4. What type of disorder is epilepsy?
   a. central nervous system
   b. metabolic
   c. respiratory
   d. circulatory

5. Seizures are related to:
   a. rheumatic fever
   b. tuberculosis
   c. epilepsy
   d. none of the above

6. Which of the following sexual activities would be least likely to spread HIV?

a. oral sex
b. vaginal intercourse
c. anal intercourse
d. masturbation

7. What type of disorder is diabetes?

a. central nervous system
b. metabolic
c. respiratory
d. circulatory

<u>Answers</u>

1. T	1. c
2. T	2. a
3. F	3. c
4. T	4. a
5. F	5. c
6. T	6. d
7. T	7. b
8. T	
9. F	
10. T	

Chapter 12 Review

For Further Consideration

We take our language and our customs for granted, and it's easy to unconsciously feel that ours is the "only way." But such an attitude is a hindrance to good communication among people of different languages and cultures. When you encounter strangers who appear to offer a barrier to communication, such as a physical disability or an unfamiliar language, it often seems that there is an invisible wall around them. However, if your desire or need to communicate is great enough (e.g., if you are lost or need a phone or restroom and must ask for help) you will see that on some level you can and do connect. If you make this effort you will always be rewarded. You will learn something new about another or about yourself. You cannot fail to be richer when you connect with a fellow human being.

Check Yourself

Terms to Define:

1. communication
2. laryngectomy
3. muteness
4. illiteracy
5. dyslexia
6. aphasia
7. articulation
8. stuttering
9. cleft palate
10. body language

True/False:

1. Language and culture are inseparable.
2. Dance is a primary means of storytelling in many countries.
3. Primitive societies frequently have simple languages.

4. Aphasia is usually caused by brain damage.

5. Illiteracy is the result of a congenital abnormality.

6. More people speak Mandarin than speak English and Spanish combined.

7. Touching is an important form of communication in all countries.

8. Esperanto is accepted as the international language.

9. Hesitant speech indicates a slowness of thought.

10. Language is a major cause of failed communication.

Multiple Choice:

1. Language allows us to:
   a. share ideas
   b. transmit culture
   c. organize society
   d. all of the above

2. A system of body language which interprets sound through movement is:
   a. eurhythmy
   b. Esperanto
   c. aphasia
   d. none of the above

3. Stuttering is a disorder of:
   a. articulation
   b. fluency
   c. vocality
   d. none of the above

4. Speaking slowly is recommended for communicating with:
   a. speakers of foreign languages
   b. illiterates
   c. people with dyslexia
   d. all of the above

5. Nonverbal communication includes:
   a. folklore
   b. pantomime
   c. intonation
   d. all of the above

6. A difficulty with pronunciation or a lisp is a problem with:
   a. articulation
   b. fluency
   c. vocality
   d. none of the above

7. Facial expressions can communicate:
   a. feelings
   b. attitudes
   c. emotions
   d. all of the above

8. Studying other languages
   a. enriches your understanding of human nature.
   b. helps you to see your culture in a universal context.
   c. improves human relations.
   d. all of the above

9. Language disorders
   a. are usually due to physical abnormalities.
   b. prevent understanding.
   c. frequently draw attention to themselves.
   d. all of the above

10. The formation of nodules or polyps may cause a disorder of:
   a. articulation
   b. fluency
   c. vocality
   d. none of the above

Answers

1. T	1. d
2. T	2. a
3. F	3. b
4. T	4. a
5. F	5. b
6. T	6. a
7. F	7. d
8. F	8. d
9. F	9. c
10. T	10. c

Chapter 13 Review

For Further Consideration

Learning more about your own unique personality type will provide insights into why other people behave the way they do. Several personality self-evaluations are available in books at the library, in which you answer a set of questions to determine your personality profile. No one personality type is better or worse than another. Every person embodies a combination of behavior styles, attitudes, and outlooks. When you are able to identify specific inborn traits governing someone's behavior, you will be empowered to tailor the situation accordingly.

The best way to handle all types of behavioral diversity is to take a course in behavior management techniques. These techniques will allow you to effectively manage situations without compromising anyone's self-esteem.

Check Yourself

Terms to Define:

1. schizophrenia
2. phobia
3. bulimia
4. ADHD
5. manic-depressive disorder
6. temperament
7. autism
8. anxiety
9. mood swings

True/False:

1. Everyone who has an emotional outburst or becomes hostile and aggressive at times has a behavior disorder.

2. The most disabling behavior disorder is considered to be psychosis.

3. A disturbance in thinking patterns is known as schizophrenia.

4. Bipolar depression involves mood swings.

5. Threatening the loss of privileges is a punishment, not a consequence.

6. Depressant medications are used to treat overactive people.

7. Mood swings may involve changes in values and attitudes.

8. Fear of heights is classified as an obsessive-compulsive disorder.

9. Tourette's Syndrome is a neurological disorder.

10. Consequence-oriented responses are appropriate for every situation.

11. ADHD is typically diagnosed by a standardized test.

Multiple Choice:

1. An anxiety attack is characterized by:
   a. fear of interacting with others
   b. negative reactions to new situation
   c. exhibiting irrationality
   d. all of the above

2. People with schizophrenia typically experience:
   a. multiple personalities
   b. hallucinations
   c. phobic disorders
   d. significant weight loss

3. Bipolar depression is characterized by:
   a. repetitive acts
   b. impulsivity
   c. poor appetite
   d. all of the above

4. Behavior changes such as high-to-low mood swings may indicate:
   a. obsessive-compulsive disorder
   b. suicidal tendencies
   c. bipolar depression
   d. both b and c

5. Which of the following symptoms is most associated with ADHD?
   a. compliance
   b. recurring fears
   c. inability to focus
   d. self-induced vomiting

6. Persons with emotional and behavioral problems may
   a. be severely antisocial and disruptive.
   b. show signs of severe anxiety or depression.
   c. vacillate between extremes of withdrawal and aggression.
   d. all of the above

7. Blurting out statements is a typical behavior of a person with
   a. depression
   b. autism
   c. ADHD
   d. phobic disorders

8. Autism is characterized by:
   a. repetitive acts
   b. impulsivity
   c. poor appetite
   d. all of the above

9. Which of the following is the best advice for dealing with disruptive behavior?
   a. avoid making eye contact
   b. respond with action
   c. prohibit the person from participating
   d. all of the above

<u>Answers</u>

1. F	1. d
2. T	2. b
3. T	3. b
4. T	4. d
5. T	5. c
6. F	6. d
7. T	7. c
8. F	8. a
9. T	9. b
10. F	
11. F	

Chapter 14 Review

For Further Consideration

Look back at the quotation that opened this chapter. George Eliot uses the word *vision* to mean *insight*, *foresight*, and *perspective*. When we have a broad perspective, we are equipped to incorporate tolerance into our daily lives. People who do not have hearing or visual disorders often have misconceptions about those who do, including beliefs that deafness and blindness lead to a life of deprived socioeconomic and cultural existence. Granted, deafness and blindness sometimes create social isolation. However, people with hearing and vision loss are capable of learning skills and enjoying leisure time and recreational activities. In reality, deafness and blindness do not hinder one from leading a successful and independent life.

Check Yourself

<u>Terms to Define:</u>

1. deafness

2. tinnitus

3. TDD

4. legal blindness

5. tunnel vision

6. hard of hearing

7. low vision

8. hearing impaired

9. visually impaired

10. partially sighted

<u>True/False:</u>

1. What hearing entails depends upon the context, the listener, and the tools he or she uses.

2. Much of what we call vision is totally subjective.

3. No one can escape the world of sound.

4. Blindness is usually correctable.

5. A radio telescope is a hearing aid.

6. A light bulb does not qualify as a visual aid.

7. Vision is the most important sense for social interactions.

8. Of all the senses, hearing is the most flexible.

9. Vision allows us to assimilate knowledge.

10. Blindness by definition limits what information an individual can obtain.

Multiple Choice:

1. We may extend our field of vision using:
   a. radio telescopes
   b. TDDs
   c. X-rays
   d. electronic cochlear implants

2. What percent of the population is considered legally blind?
   a. .1%
   b. 1%
   c. 10%
   d. 20%

3. A person with profound hearing loss may be able to hear:
   a. loud environmental sounds
   b. loud voices
   c. both a and b
   d. none of the above

4. Sound waves are collected by the:
   a. auditory canal
   b. eardrum
   c. outer ear
   d. cochlea

5. Light is initially captured by the:
   a. retina
   b. optic nerve
   c. cornea
   d. pupil

6. The term *visually impaired* may include individuals with:
   a. partial sight
   b. complete loss of sight
   c. prenatal blindness
   d. all of the above

7. The term *hearing impairment* may include:
   a. mild to moderate loss
   b. moderate to severe loss
   c. severe to profound loss
   d. all of the above

8. If a totally blind person is exposed to bright light, it will affect his or her:
   a. moods
   b. hormones
   c. both a and b
   d. none of the above

9. Which of the following is *not* an element of blind culture?

a. getting someone's attention
b. leave-taking
c. introducing oneself
d. none of the above

Answers

1. T	1. c
2. T	2. a
3. T	3. a
4. F	4. c
5. T	5. a
6. F	6. d
7. F	7. d
8. T	8. c
9. T	9. d
10. F	

Chapter 15 Review

For Further Consideration

We have seen that the definition of family is broad and varied. Everyone belongs to various family structures and social circles which may overlap.

Families constituted by relatives provide us with nurturing, love, support, trust, shelter, food, attention, worth, and dignity. Families constituted by friends and peers help us to explore a broader world, interact with others, appreciate diversity, and enjoy sharing our lives. Families constituted by co-workers and employers help us learn to cooperate, complete tasks, share the workload, offer assistance, and solve problems.

Check Yourself

Terms to Define:

1. child abuse

2. circle of friends

3. migrant family

4. latchkey kids

5. extended family

6. emotional abuse

7. domestic abuse

8. neglect

9. nontraditional family

10. nuclear family

True/False:

1. No family can escape having members who fit some category of exceptionality.

2. The typical American family now has a single parent as head of household.

3. The most important learning environment for any child is the family.

4. Children who grow up with uneducated parents are no less likely to succeed than children who grow up with educated parents.

5. A nuclear family is made up of several generations.

6. Nuclear families make up approximately 50% of American families.

7. Sexual abuse by definition involves physical contact.

8. One in three or four girls will be sexually abused by age eighteen.

Multiple Choice:

1. Which of the following does not qualify as a "normal" family?
   a. single-parent household
   b. reorganized family
   c. extended family
   d. none of the above

2. Which of the following families is most likely to be happy and successful?
   a. nuclear family
   b. extended family
   c. step family
   d. none of the above

3. Which of the following families is created through marriage, remarriage, or cohabitation?
   a. reorganized family
   b. extended family
   c. nuclear family
   d. migrant family

4. Which of the following families consists of several generations living in the home?
   a. reorganized family
   b. extended family
   c. nuclear family
   d. none of the above

5. Which of the following constitutes an "ideal" family?
   a. father, mother, two children
   b. grandparents and other relatives living at home
   c. support system of friends and neighbors
   d. opinions vary according to individual beliefs, societal trends, and cultural traditions

6. The process of accepting an exceptional family member may involve:
   a. denial, guilt, anger
   b. adjustment, education, acceptance
   c. both a and b
   d. none of the above

7. Children who spend longer than three hours at home alone every day are called
   a. throwaways
   b. latchkey kids
   c. abandoned children
   d. none of the above

8. Which of the following families is characterized by relocation to a new country?

a. reorganized family
b. extended family
c. nuclear family
d. migrant family

Answers

1. T	1. d
2. F	2. d
3. T	3. a
4. F	4. b
5. F	5. d
6. F	6. c
7. F	7. b
8. T	8. d

Chapter 16 Review

For Further Consideration

Human beings have an innate thirst for knowledge. It is through education that we cultivate our minds to enable us to accomplish all we would like to accomplish in life. Education enables us to make full use of our potential. All around the world, educational systems and curricula are being reformed to accommodate new trends in society. Whether one is entirely self-taught or attends public, private, or parochial schools, the goal of education should be to fully develop an individual into a responsible citizen of the world. Such a citizen understands and appreciates the full range of human thought, recognizing that truth and knowledge may take many different forms and may be approached from a multiplicity of perspectives.

Check Yourself

Terms to Define:

1. risk factors
2. public schooling
3. home schooling
4. parochial schooling
5. special education
6. Individualized Education Plan
7. zero reject
8. mainstreaming
9. inclusion

True/False:

1. A private school offers an environment which is likely to be culturally and racially diverse.

2. Children taught at home by their parents are likely to score below average on achievement tests.

3. Special education is designed to help students achieve personal self-sufficiency and academic success.

4. According to Public Law 94-142, children and youth with disabilities should rarely be placed in general education classes so that their special needs may be fully addressed.

5. Students have more needs in common than they have differences.

Multiple Choice:

1. A child's individualized education plan is designed by:
   a. the child
   b. the parent or guardian
   c. the teacher
   d. all of the above

2. The concept that children and youth with disabilities should be educated alongside nondisabled students to the maximum extend possible is known as:
   a. nondiscriminatory evaluation
   b. least restrictive environment
   c. due process
   d. zero reject

3. Which type of schooling specializes in offering social interaction with a broad range of students?
   a. public
   b. private
   c. parochial
   d. alternative

4. Which type of schooling specializes in offering flexible schedules and intensive studies?
   a. public
   b. private
   c. parochial
   d. alternative

5. Which type of schooling is based upon an individual or traditional learning philosophy?
   a. public
   b. private/parochial
   c. special
   d. alternative

6. Which type of schooling is best?
   a. public
   b. private/parochial
   c. alternative
   d. depends upon the individual

Answers

1. F	1. d
2. F	2. b
3. T	3. a
4. F	4. d
5. T	5. b
	6. d

Chapter 17 Review

For Further Consideration

As you end your study of human diversity, here is a final challenge. You are now educated about the varieties of human differences, and you will notice that many of your friends, family

members, and fellow members of society are still uninformed about diversity. Your challenge is to avoid letting your ego rule your actions. It is easy to feel superior to people whose comfort level differs from yours, or whose attitudes seem less enlightened than yours. Show *everyone* the same compassion and understanding. Remember that we are *all* exceptional people. We are all at different stages of development, yet everyone is learning every day. Teach others by your own example, and always be aware of practicing respect.

Check Yourself

True/False:

1. Often merely having the intention to be comfortable will help you to actually feel comfortable.

2. It is appropriate to ask persons who are diverse questions about issues in which you are interested.

3. In order to show your respect for different labels, it is important to treat some people differently than you would anyone else.

4. If someone is not comfortable with your diverse friends, the response may be due to ignorance, and you can use that as an opportunity to explain your perspective.

5. Be careful to adjust your vocabulary around people who are different.

6. Try your best to seem pleasant and happy around those who are different, even if you are in a bad mood.

7. Because everyone is different, no one needs special accommodations.

8. People who are different deserve every opportunity to lead the normal life they desire.

9. It is enough to tolerate people who are different, even if you don't respect them as important members of society.

Answers

1. T
2. T
3. F
4. T
5. F
6. F
7. F
8. T
9. F

Notes

Index

D

E

J

K

L

M

S

T

U

V

W

Y

Z